USBORNE

WEATHER & CLIMATE CHANGE

USBORNE

WEATHER & CLIMATE CHANGE

Laura Howell

Designed by Joanne Kirkby,
Karen Tomlins and Laura Hammonds

Digital imagery and illustrations by Keith Furnival

Edited by Kirsteen Rogers

Cover design: Michael Hill

Consultant: Dr Roger Trend, University of Exeter

Website adviser: Lisa Watts

Internet links

Throughout this book, we have suggested interesting websites where you can find out more about weather and climate change. To visit the sites, go to the Usborne Quicklinks Website at **www.usborne-quicklinks.com** and type the keywords "science weather". There you will find links to click on to take you to all the sites. Here are some of the things you can do on the websites:

- make your own weather station and become a weather forecaster in your own home

- become a climate detective and find all the latest news and views on global warming

- watch animated movies and take quizzes about many weather and climate change topics

Notes for parents

The websites described in this book are regularly reviewed and the links in Usborne Quicklinks are updated. However, the content of a website may change at any time and Usborne Publishing is not responsible for the content of any website other than its own.

We recommend that children are supervised while on the Internet, that they do not use Internet chat rooms, and that you use Internet filtering software to block unsuitable material. Please ensure that your children read and follow the safety guidelines printed on the right. For more information, see the "Net Help" area on the Usborne Quicklinks Website.

See for yourself

The *See for yourself* boxes in this book contain experiments or activities that we have tested. Some recommended websites also contain experiments, but we have not tested all of these. This book will be used by readers of different ages and abilities, so it's important that you do not tackle any experiment involving equipment that you wouldn't normally use, such as a knife. Instead, ask an adult to help you.

Internet safety

When using the Internet, please make sure you follow these guidelines:

- Ask your parent's or guardian's permission before you connect to the Internet.

- If you write a message in a website guest book or on a website message board, do not include any personal information such as your full name, address or telephone number, and ask an adult before you give your email address.

- If a website asks you to log in or register by typing your name or email address, ask permission of an adult first.

- If you receive an email from someone you don't know, tell an adult and do not reply to the email.

- Never arrange to meet anyone you have talked to on the Internet.

Downloadable pictures

Pictures in this book marked with a ★ symbol can be downloaded from Usborne Quicklinks for your own personal use, for example, to illustrate a homework report or project. The pictures are the copyright of Usborne Publishing and may not be used for any commercial or profit-related purpose. To download a picture, go to Usborne Quicklinks and follow the instructions there.

Site availability

The links in Usborne Quicklinks are regularly reviewed and updated, but occasionally you may get a message that a site is unavailable. This might be temporary, so try again later, or even the next day. Websites do occasionally close down and when this happens, we will replace them. Sometimes we add extra links too, so when you visit Usborne Quicklinks, the links may be slightly different from those described in your book.

Computer not essential

If you don't have access to the Internet, don't worry. This book is a complete, self-contained reference book on its own.

Contents

These beautiful shapes are clusters of snowflakes seen up close.

What makes the weather?

All of Earth's many kinds of weather depend on two things: water and the Sun. Water moves between the oceans, skies and land, forming clouds and all kinds of wet weather. The Sun heats the Earth and causes air to move from place to place. We feel this air movement as winds. Without these different types of weather, no life could exist on Earth.

Earth and its weather

Weather, such as rain, wind or snow, happens all around us every day. A pattern of weather that repeats itself over a long time is described as climate. So, a place with lots of rain year after year has a wet climate. Earth is the only planet with just the right types of climate for living things; all the others are too hot or cold.

Changing weather

In many places on Earth, weather changes daily. Weather scientists called meteorologists try to predict what the weather will be like in the future, either from day to day, or over a longer period. They do this by recording and measuring temperature, the amount of moisture in the air, wind speed and many other things.

Venus (hotter than Earth)

Sun

Earth

Distance from the Sun makes a difference to a planet's weather and climate.

Mars (colder than Earth)

Weather satellites like this fly around Earth, gathering information about winds, moisture, clouds and the temperature of the atmosphere.

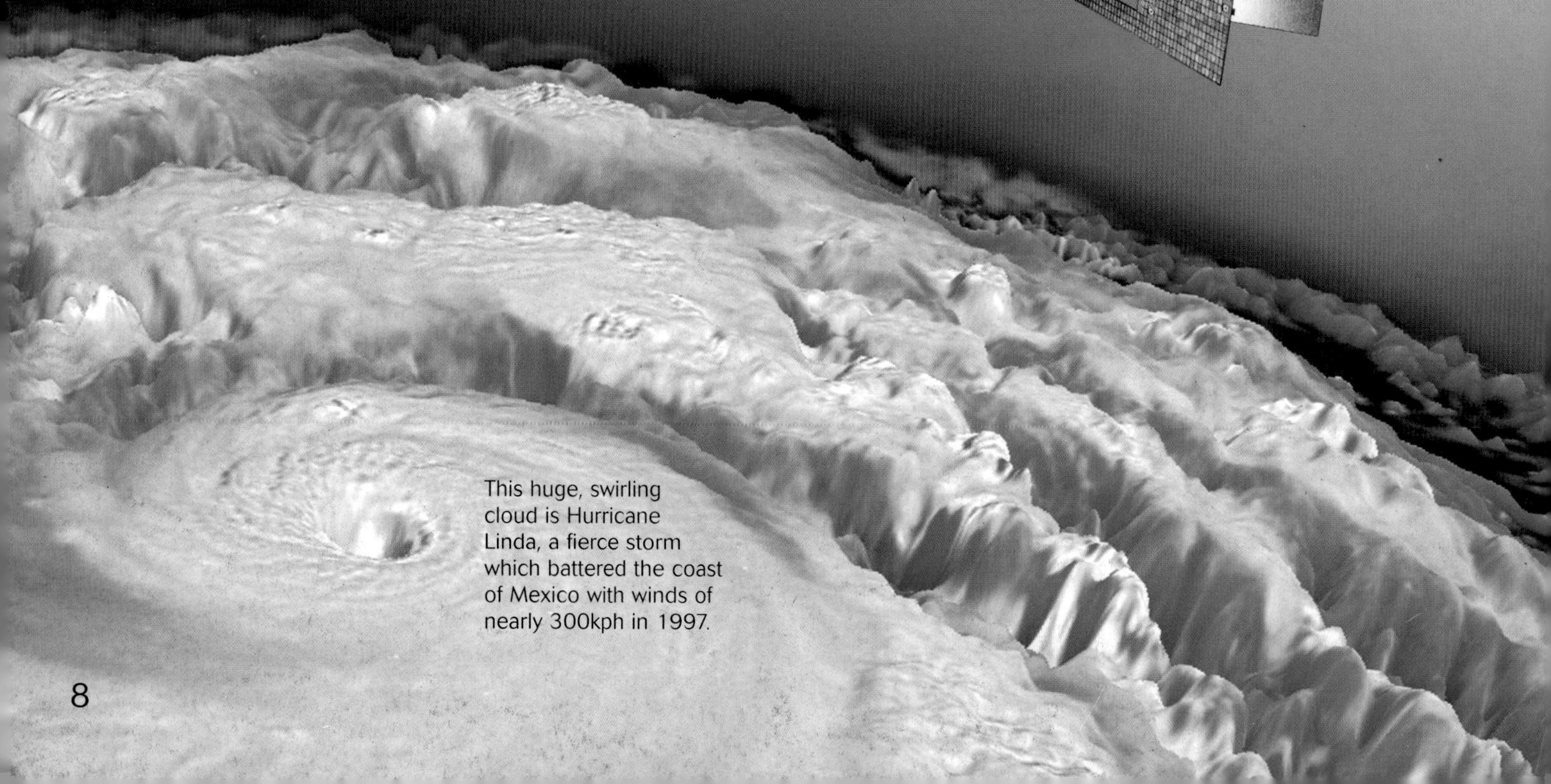

This huge, swirling cloud is Hurricane Linda, a fierce storm which battered the coast of Mexico with winds of nearly 300kph in 1997.

The climate system

It may seem as if weather comes from nowhere, but many things work together to create it. This huge "weather machine" is called the climate system. Air, ice, water, energy from the Sun, and even living things all play a part.

Moving heat

The main thing the climate system does is move heat between water, air and land. This is called the global energy system. If Earth's heat wasn't spread around, its hot parts would be hotter and its cold parts colder.

★

The Sun heats the oceans and land.

Sun

The atmosphere keeps Earth from becoming too hot or cold.

Tiny water drops rise and make clouds. Water falls as rain, snow or hail.

Clouds

Earth gives out heat as the Sun warms it up.

Ice sheets, snow and sea ice reflect sunlight.

Land

Oceans

Plants and animals add gases to the atmosphere.

Weather and water

No other known planet has weather that changes as much as ours. This is mainly because Earth is the only one with water on its surface. The air and seas exchange water and heat all the time, creating clouds, wind and all sorts of weather. We call all the water on or around Earth the hydrosphere.

About 71% of Earth's surface is covered in water. Almost half is in the vast Pacific Ocean, shown here.

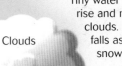

INTERNET LINKS

For a link to the following website, go to **www.usborne-quicklinks.com** and enter the keywords "science weather".

Website 1 Take an interactive look at climate and weather, with animations.

A varied climate

Earth's climate is not the same over its whole surface. It is also constantly changing. Usually this happens very slowly, over tens of thousands of years. It's possible, though, that humans are having a hand in climate change today, and speeding it up.

Extreme temperatures

The places on Earth with the most extreme temperatures are the equator – an imaginary line around the planet's middle – and the poles, at Earth's "top" and "bottom". Equatorial areas are hot all year round, but the poles are always cold, even in summer.

The computer image on the right shows Earth's hottest areas in red and its coldest areas in blue.

INTERNET LINKS

For links to the following websites, go to **www.usborne-quicklinks.com** and enter the keywords "science weather".

Website 1 Compare climates around the world with an interactive map.

Website 2 Satellite images and information about our amazing Earth.

There is a huge variety of living things on Earth because of its varied climate. These flowers, for instance, only grow in certain rainforests near the equator, where it is hot and damp enough for them.

North Pole

Equator

South Pole

The landscape below is part of Antarctica, a vast, frozen continent at the South Pole. It is colder than anywhere else on Earth.

Climate change in the past

Earth's climate has been changing ever since the planet formed. Scientists think that to start with, Earth was very hot, and that it cooled as the atmosphere slowly formed over millions of years. There have been periods of warming and cooling ever since.

Earth was once home to huge animals like this monstrous sea reptile. Around 65 million years ago, these animals died out, probably due to a change in the climate, which they could not survive.

Getting warmer...

In the last century or so, the climate seems to have changed unusually quickly. The planet's average temperature is rising, with all kinds of alarming effects. Many people think that human activities are mostly to blame.

Global warming may be causing a rise in sea levels. This leads to more flooding of coastal areas and islands, such as Zanzibar, above.

What's really happening?

The climate system is so huge and complex that changes are difficult to identify. It's even harder to know exactly what causes them. Climate experts disagree about what is happening right now, let alone what might happen in the future.

Some experts think humans are causing terrible damage to the climate system, which can't be fixed. Others say human activities won't affect climate at all. Many think we don't know enough to say for certain either way.

See for yourself

Both water and land absorb heat from the Sun, but at different rates. You can feel this effect if you go to the beach on a hot day. Although the sand may be too hot to walk on, the sea water is still cool enough to swim in.

The Sun

The Sun is our great life-giver – nothing could survive on Earth without its heat and light. But it has a deadly power too, which the climate system helps keep at bay.

Energy from the Sun

The Sun gives out rays, called solar radiation. The rays travel in waves of energy. Waves with short wavelengths have more energy than waves with long wavelengths.

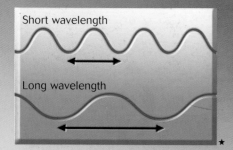

Short wavelength

Long wavelength

★

The shorter the wavelength, the more energy the wave has.

There are three main types of solar radiation.

• Infrared (IR) rays have a very long wavelength. They are low-energy rays which heat things up.

• Visible light rays allow us to see things. They look colourless, but actually contain all the colours of the rainbow mixed up.

• Ultraviolet (UV) rays are high-energy, with a short wavelength. These rays can damage living things: for instance, they can cause skin cancer.

Solar panels like the one on this roof soak up the Sun's energy and use it to provide power in the house.

Some birds, like this flock of geese, use the Sun to find their way.

There are different amounts of infrared, ultraviolet and visible light rays in solar radiation.

★

Infrared rays
49%

Visible light rays
44%

Ultraviolet rays
7%

Colours in the sky

The atmosphere contains millions of tiny particles: for instance, dust, water vapour and gas molecules.

As visible light rays travel through the atmosphere, they hit particles and are bent, split up and scattered in different directions. This is because each colour of light has a different wavelength. The way that light scatters makes the sky appear to turn different shades.

Clear blue skies

On a clear day, the sky looks blue. Light rays travel through the atmosphere and are scattered. Blue, indigo and violet, the colours with the shortest wavelengths, hit the most particles and are widely scattered. We don't see indigo and violet, so the sky looks blue.

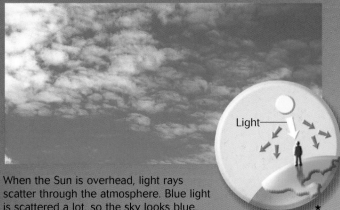

When the Sun is overhead, light rays scatter through the atmosphere. Blue light is scattered a lot, so the sky looks blue.

Red sky at night...

In the morning and evening, the Sun is lower in the sky, so its rays have to travel through more of the atmosphere. Red and orange light have the longest wavelengths, and pass through the particles more easily than any others. These colours reach your eyes, and the sky looks red.

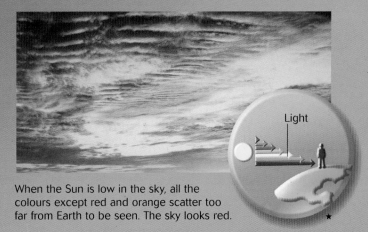

When the Sun is low in the sky, all the colours except red and orange scatter too far from Earth to be seen. The sky looks red.

The seasons

Many places on Earth have four seasons – spring, summer, autumn and winter – which are caused by Earth being tilted at an angle to the Sun. As it makes its year-long journey around the Sun, parts of Earth are warmed by different amounts. This causes the weather to change regularly throughout the year.

Plants need sunlight to make food. Sunflowers turn to face the Sun, so they get as much sunlight as possible.

How the seasons change

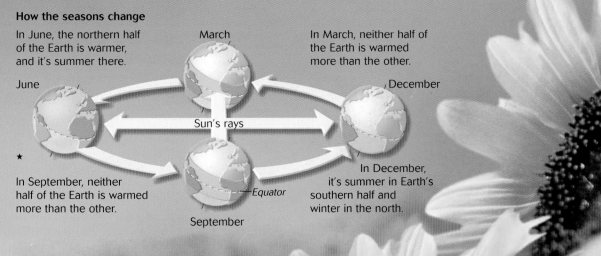

In June, the northern half of the Earth is warmer, and it's summer there.

June

In September, neither half of the Earth is warmed more than the other.

March

September

Equator

Sun's rays

In March, neither half of the Earth is warmed more than the other.

December

In December, it's summer in Earth's southern half and winter in the north.

Soaking up the sun

Dark areas, such as soil, absorb the Sun's heat and warm up more quickly than light areas such as ice. Water absorbs heat more slowly than land, but stores it for longer.

How much heat an area receives from the Sun depends on whether the Sun is overhead or not. Near the equator, the Sun is almost directly overhead. The rays are more concentrated than elsewhere, creating a hot climate.

These oranges only grow in places where the climate is warm and sunny for most of the year.

INTERNET LINKS

For links to the following websites, go to www.usborne-quicklinks.com and enter the keywords "science weather".

Website 1 Take a trip through the Sun, with amazing facts and movies.

Website 2 Explore the nature of the Sun, with video clips and images.

Website 3 A really useful site with lots of information about the Sun.

Near the equator the Sun is almost overhead. The rays feel strong. ★

Equator

Sun's rays

At the poles, the Sun is not overhead. The rays are spread over a wide area and bounce off ice sheets. Temperatures are almost always below 0°C.

What happens to the Sun's rays

Different things happen to the Sun's rays as they reach Earth.

Sun

Some are reflected back into space by clouds.

Some are absorbed by clouds.

Some clouds reflect heat back to Earth.

As Earth is warmed, it heats the air above it.

Some bounce off clouds, then down to Earth.

Most are absorbed by Earth's surface. ★

At the poles, the Sun is never overhead. The rays feel weak. ★

Sun's rays

South Pole

Spots on the surface

The Sun's surface has huge, cooler patches on it called sunspots. These grow and shrink in cycles lasting roughly 11 years. As they change in size, the amount of energy coming from the Sun changes too. Some experts think this might be partly to blame for recent climate change.

The dark areas on this image of the Sun are sunspots – cooler patches on its surface.

See for yourself

Looking directly at the Sun can blind you, but here is a safe way for you to see sunspots.

Point a pair of binoculars at the Sun with a piece of white cardboard behind them but DON'T LOOK THROUGH THEM. Move the binoculars around until a white circle appears on the cardboard. This is the Sun's image. Focus the binoculars until the image is sharp. The dark smudges on the image are sunspots.

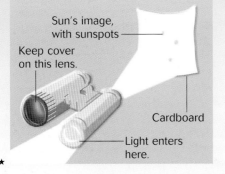

Sun's image, with sunspots

Keep cover on this lens.

Cardboard

Light enters here.

Energy from the Sun

For about 20 years, scientists have studied and measured the amount of energy given out by the Sun. Recently, they have begun to understand more about how it's linked to climate.

High-energy rays from space constantly stream towards the Earth. If the Sun gives out less energy, more of these rays can reach the planet. This seems to cause more clouds to form high up in the air. Clouds can both add to and reduce the effects of global warming.

Global warming leads to melting ice in Antarctica.

Getting hotter

Right now, Earth's average surface temperature is about 15°C. If global warming continues, this could rise by between 1.5 and 4.5 degrees in the next hundred years. This would make Earth's average temperature higher than it's been for several million years.

The atmosphere

Earth is surrounded by the atmosphere – a blanket of gases which protects us from the Sun's harmful rays and stops Earth from getting too hot or cold. All of Earth's weather happens in the lowest 10km of the atmosphere.

Atmospheric layers

Earth's atmosphere is divided into layers, according to the temperature of the gases in it. Each layer has a name ending in "sphere", and the boundaries between the layers have names which end with "pause". The atmosphere is held in place by gravity, a force which attracts objects to each other. Gases in the atmosphere are attracted to Earth.

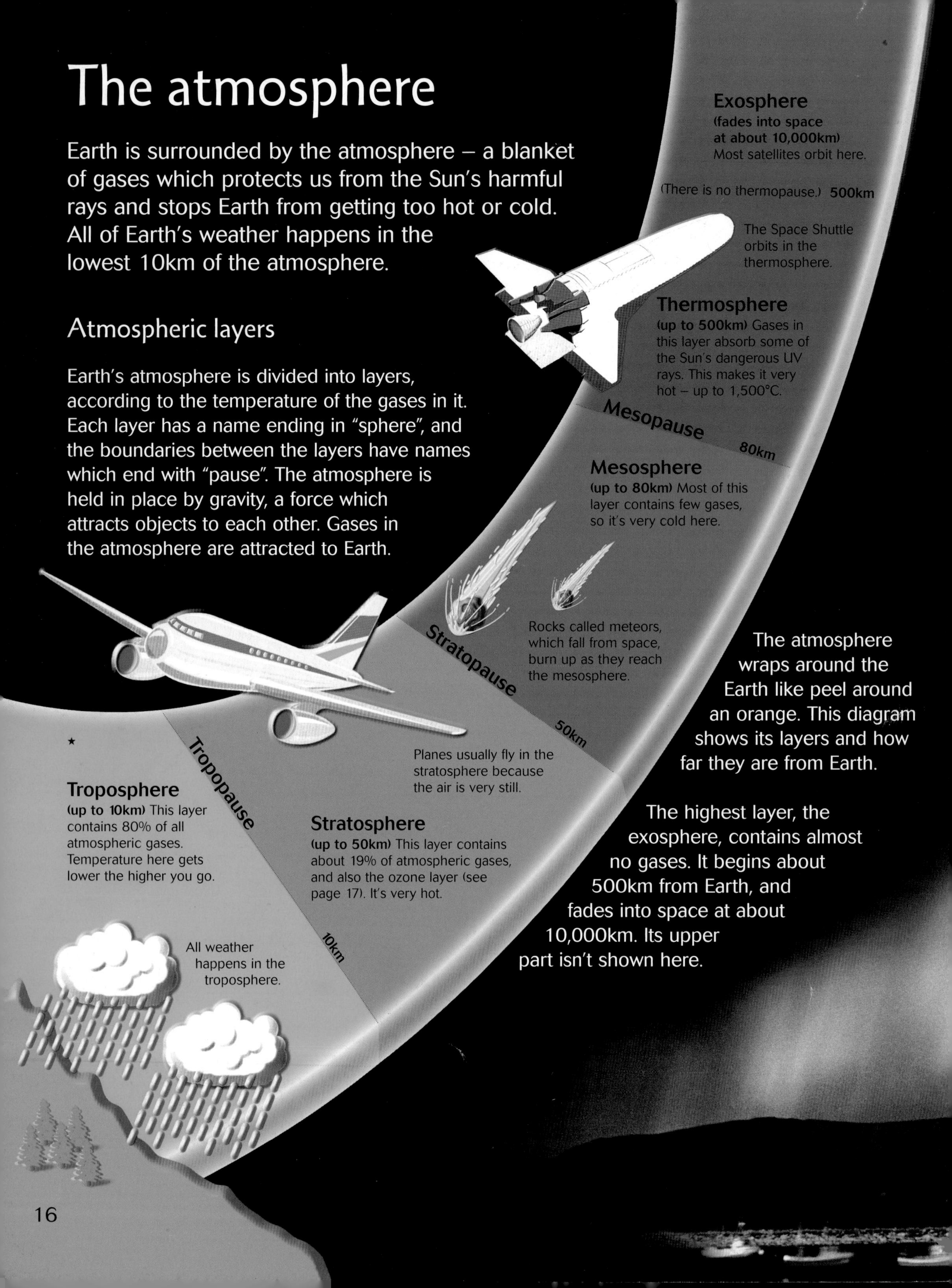

Exosphere
(fades into space at about 10,000km) Most satellites orbit here.

(There is no thermopause.) **500km**

The Space Shuttle orbits in the thermosphere.

Thermosphere
(up to 500km) Gases in this layer absorb some of the Sun's dangerous UV rays. This makes it very hot – up to 1,500°C.

Mesopause

80km

Mesosphere
(up to 80km) Most of this layer contains few gases, so it's very cold here.

Stratopause

Rocks called meteors, which fall from space, burn up as they reach the mesosphere.

50km

The atmosphere wraps around the Earth like peel around an orange. This diagram shows its layers and how far they are from Earth.

The highest layer, the exosphere, contains almost no gases. It begins about 500km from Earth, and fades into space at about 10,000km. Its upper part isn't shown here.

Tropopause

Planes usually fly in the stratosphere because the air is very still.

★

Troposphere
(up to 10km) This layer contains 80% of all atmospheric gases. Temperature here gets lower the higher you go.

Stratosphere
(up to 50km) This layer contains about 19% of atmospheric gases, and also the ozone layer (see page 17). It's very hot.

10km

All weather happens in the troposphere.

Atmospheric gases

The atmosphere is made of many gases, although some exist in very tiny amounts. As well as the gases listed below, the atmosphere contains water in three forms: a gas, called vapour, ice crystals and droplets.

Percentage of gases in the atmosphere

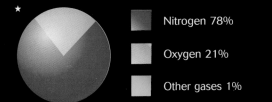

Nitrogen 78%

Oxygen 21%

Other gases 1%

The 1% of other gases includes argon, neon, krypton and xenon, and harmful chemicals such as nitrogen dioxide.

See for yourself

The atmosphere pushes down on Earth with a force called air pressure or atmospheric pressure. To see this force in action, put a deflated balloon under a small book on a table and blow hard into the balloon. The book is lifted by the pressure of the air you blow into the balloon.

INTERNET LINKS

For links to the following websites, go to **www.usborne-quicklinks.com** and enter the keywords "science weather".

Website 1 Atmosphere diagrams and facts.

Website 2 Take a self-guided tour of auroras, with fantastic pictures, sounds and movies.

The ozone layer

Ozone is a gas which exists as a layer in the stratosphere. It absorbs high-energy UV rays from the Sun, protecting us from them. But over the last few decades, this natural shield has been damaged by man-made chemicals called CFCs. At certain times of year, a hole appears over the South Pole, and in other areas the layer becomes very thin. This lets in more UV rays.

In recent years, several countries have made agreements to cut the use of CFCs, and the hole is shrinking. Even so, experts think it will take at least 50 years to repair itself.

The ozone hole (shown in purple) is a little different every year. In this picture, from 2002, it's almost split in two.

Antarctica

Lights in the sky

Around the poles, tiny electrically charged particles high in the atmosphere react with rays from the Sun to create beautiful sheets of light. These are called the Aurora Borealis or northern lights (near the North Pole) and the Aurora Australis or southern lights (near the South Pole). Auroras can appear in many different shapes and colours.

This photo shows a rare red glow of light, called an aurora, over Skarsvag in Norway.

Greenhouse gases

A tiny number of gases in the atmosphere are able to trap heat around Earth. Without them, heat would be lost into space and we'd all freeze. The trapping of heat around Earth is called the greenhouse effect, because glass in a greenhouse traps heat in a similar way, keeping the inside of the greenhouse warm.

Too much of a good thing?

Greenhouse gases are like a blanket: if they are the right thickness, Earth won't be too hot or too cold. That's the way Earth has been for hundreds of thousands of years, because there has been just the right amount of greenhouse gases.

Many experts now think the way we burn fuel and add chemicals to the atmosphere is causing the blanket to become thicker. This makes Earth hotter, an effect known as global warming.

In the Eden Project building in Cornwall, greenhouse glass makes it warm enough to grow tropical plants. It's too cold outside.

Sun's rays

Greenhouse gases

Heat from Earth is absorbed by greenhouse gases and then sent back again.

Greenhouse gases work like glass in a greenhouse. They let heat from the Sun through, but send back heat that Earth gives out.

Water vapour

Water vapour is Earth's most common greenhouse gas: it's responsible for about 60% of the natural greenhouse effect. Although human activities have very little direct effect on water vapour levels, global warming is causing them to increase.

Heated water from oceans, rivers and lakes turns to water vapour. Global warming makes more vapour form.

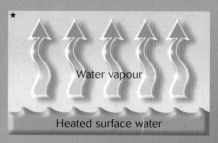

Water vapour

Heated surface water

INTERNET LINKS

For links to the following websites, go to **www.usborne-quicklinks.com** and enter the keywords "science weather".

Website 1 A guide to greenhouse gases and global warming.

Website 2 Find out more about greenhouse gases, where they come from and how they keep the Earth warm.

Carbon dioxide

Carbon dioxide (CO_2) enters the air in many ways: for instance, when plants decay, volcanoes erupt and animals breathe. When we burn fossil fuels, such as oil, this also releases huge amounts of CO_2 into the atmosphere. Carbon dioxide is thought to be the main gas responsible for global warming.

Power plants such as this add large amounts of carbon dioxide and water vapour to the air as they burn fuel.

Methane

Methane can trap about 21 times more heat than the same amount of carbon dioxide. It's given out by animals as a waste gas, by rotting plants and rubbish, and by burning fossil fuels.

Although there is much less methane than CO_2 in the atmosphere, levels are rising by about 0.6% per year, the largest increase in any natural gas. Most of this is produced as a result of human activity.

Decaying plants in rice paddies like these release methane. But rice is an essential food crop, so we can't stop growing it.

Nitrous oxide

Nitrous oxide is the most powerful greenhouse gas – it can trap around 270 times more heat than the same amount of CO_2. Luckily, it only exists in the atmosphere in tiny amounts.

Ozone

Ozone is a helpful gas in the stratosphere, but it's dangerous in the troposphere. It forms there when sunlight reacts with exhaust fumes from cars and trucks. This ozone acts as a powerful greenhouse gas, and is also harmful to living things.

19

The greenhouse effect

The greenhouse effect is the way that greenhouse gases, such as carbon dioxide, help to heat up a planet. It happens naturally on every planet with an atmosphere, but to different extents.

The greenhouse effect is caused by the Sun's heat being trapped around Earth.

The greenhouse effect on Earth

Not all the heat we feel on Earth comes straight from the Sun. Most of it is heat from Earth itself, which is trapped and sent back down by greenhouse gases.

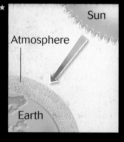
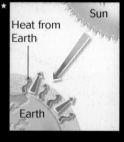

Greenhouse gases let in high-energy radiation from the Sun. This warms Earth, but not the atmosphere.

As Earth warms up, it gives out heat as low-energy infrared rays. Only a few of these pass through the greenhouse gases.

The greenhouse gases trap most of the infrared rays. This heats the atmosphere, which in turn heats Earth.

Venus and Earth

Life on Earth depends on the greenhouse effect, but conditions on other planets show it can be harmful too.

Venus, for instance, has many similarities to Earth, such as its size and distance from the Sun. But, unlike Earth, it has a deadly average temperature of 470°C.

This is mainly due to its atmosphere, which is 96% carbon dioxide. Earth's is only 0.03%. This means that far more heat is trapped around Venus than Earth.

INTERNET LINKS

For links to the following websites, go to **www.usborne-quicklinks.com** and enter the keywords "science weather".

Website 1 An animated explanation of the greenhouse effect.

Website 2 Read more about the causes of the greenhouse effect.

These clouds around Earth formed from water vapour, a greenhouse gas. Without greenhouse gases to warm it up, Earth would be a bleak, frozen ball.

Enhanced effect

Without greenhouse gases, Earth's average surface temperature would be around −18°C, which is too cold for life. Natural greenhouse gas levels make it a comfortable 15.5°C. But when more greenhouse gases are added, for instance by human activities, the greenhouse effect is enhanced, and global temperatures start to rise.

What's the cause?

The main gas that scientists have linked to an enhanced greenhouse effect is carbon dioxide (CO_2). Levels of CO_2 in the atmosphere are currently at their highest for 160,000 years.

It's thought that around a quarter of the CO_2 in the atmosphere today was added by human actions. The rest comes from the natural world.

Greenhouse gases and nature

Natural processes such as breathing release ten times more carbon dioxide than human activities such as burning fossil fuels. But until recently, carbon dioxide levels remained balanced by sinks – areas such as forests and oceans which absorb carbon. There is now too much carbon dioxide for sinks to absorb.

Carbon dioxide increase

Since the end of the 18th century, the amount of carbon dioxide in Earth's atmosphere has increased from about 280 parts of carbon dioxide per million parts of air, to about 370 parts per million. It's continuing to rise at a rate of about 1.2 parts per million each year.

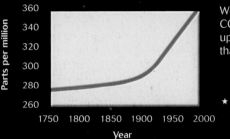

Within 100 years, CO_2 levels could be up to 150% higher than today's levels.

Methane increase

Carbon dioxide is not the only greenhouse gas which is on the increase. During the same period as above, methane levels have more than doubled, to 1.75 parts per million. Although this doesn't sound much, it's thought to have enhanced the greenhouse effect by about 20%.

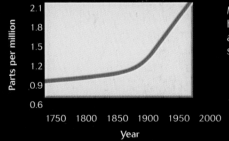

Methane levels have risen by around 30% since the 1950s.

Most living things give out greenhouse gases in some form. A single cow releases around 600 litres of methane a day, as it digests its food.

Global warming

When the greenhouse effect gets stronger, global warming takes place — that is, Earth's average temperature rises. Then, conditions on Earth start to change. It's important for us to understand how and why this happens, because more and more evidence suggests that we are to blame for it.

Where's the proof?

Many people still argue about the reality of global warming, but there's a lot of evidence to support it. For instance, various studies show that:

• global surface temperatures have increased by between 0.4 and 0.8°C since the late 19th century.

• the 20th century's ten warmest years all happened after 1985. Of these, 1998 was the warmest year on record, and 2001 the second.

• snow, mountain glaciers and Arctic sea ice in the northern hemisphere are all melting, causing sea levels to rise.

• sea levels have risen by around 10-20cm over the past century.

• rainfall around the world has risen by about 1%. More water vapour in the air is causing heavy rain in more areas.

Studying global warming

Many organizations study climate change and global warming, but the main one is the IPCC (Intergovernmental Panel on Climate Change). This is made up of over 3,000 scientists from different nations, who gather and assess climate information from around the world.

About 25,000 vehicles cross the Golden Gate Bridge in San Francisco, USA every day. By 2010, there may be over 800 million motor vehicles in the world, and most of them will produce gases which add to global warming.

INTERNET LINKS

For links to the following websites, go to **www.usborne-quicklinks.com** and enter the keywords "science weather".

Website 1 See the effects of global warming with a clickable world map.

Website 2 A clickable guide to climate change, with a quiz.

The IPCC's findings

The IPCC produces regular reports of what its collected data shows about climate change. Their 2001 report claimed there's very strong evidence that human activities affect the atmosphere. Not only that, but more studies are backing up the idea that humans are almost entirely to blame for global warming over the last 50 years.

Future warming

The IPCC also predicts that the pattern of global warming which began in the last century will continue through this one. This is likely to have many damaging effects on both the planet and its people, especially in places that have already been weakened by famine and poverty.

Temperatures are rising in the South American forests where toucans live. The birds are forced to move to cooler areas in order to survive.

Many Indian states, such as Orissa, have suffered severe flooding in recent years. If global warming continues, the situation is likely to get worse, as floods bring disease and kill crops.

How much hotter?

Over the next century, Earth's average surface temperature could rise faster than at any time in the last 10,000 years. Scientists expect an increase of between 0.6 and 2.5 degrees in the next fifty years, and between 1.4 and 5.8 degrees in the next century. The amount of warming will vary from place to place.

Where's the warming?

The maximum warming will probably be in the Arctic, in winter. Arctic sea ice has already thinned by 40% since the 1970s; it's possible that Greenland's land ice may even disappear.

The effects aren't confined to cold areas, though. Many milder countries, such as the UK, are experiencing changes in the seasons, with unusually warm winters and wet summers. Spring is also thought to be getting earlier each decade.

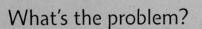

Some animals are adapting to earlier springtimes. Red squirrels, for instance, are having their babies earlier in the year.

What's the problem?

Sometimes, people don't understand why global warming might be a problem – after all, in many areas it will create a warmer and more pleasant climate. But experts predict it will drastically change the world as we know it.

Climate zones

The land surface of the Earth can be divided into different zones, or biomes, based on climate. Each type of climate zone has its own unique weather, landscape and living things. As climate change takes place, conditions in these zones change too.

Map showing main world biomes described on pages 24–29

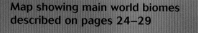

- Tropical forest
- Tropical grassland
- Desert
- Dry grassland
- Warm temperate (Mediterranean)
- Cool temperate
- Boreal forests (cold temperate)
- Tundra
- Mountain
- Polar area

(On a flat map, Antarctica is usually shown spread out because of its position at the bottom of the globe.)

Save the rainforest!

Rainforests are tropical forests near the equator. They are full of living things, and rich in resources. But many rainforests are being damaged or destroyed by mining, farming, logging and clearing land to meet growing human needs. Visit the websites described on the next page to learn about the threats to rainforests.

Most frogs live in water, but these tree frogs live in rainforest trees. The climate there is so moist that they get all the water they need from the air.

The Tropics

The tropics are two imaginary lines which circle the Earth. The Tropic of Cancer is at 23°27′ north, and the Tropic of Capricorn is at 23°27′ south. The areas that lie between these two lines are described as "the Tropics". They have hot climates, which are usually described as tropical.

Location of the Tropics

Tropical forests

Tropical forests are very hot, with temperatures usually between 26 and 28°C.

The forests furthest from the equator, called tropical moist forests, have two wet and two dry seasons. Forests around the equator, called rainforests, have no seasons at all, just rain all year round. There are also regular thunderstorms. These are caused by the land heating up and making moist forest air rise and form stormclouds.

INTERNET LINKS

For links to the following websites, go to www.usborne-quicklinks.com and enter the keywords "science weather".

Website 1 Facts and figures about each different biome, with activities too.

Website 2 An interactive guide to forests and climate change.

Forests in danger

Trees act like natural sponges, soaking up carbon dioxide from the air. This helps prevent global warming. Areas such as forests, which can store lots of carbon, are called carbon sinks.

Huge areas of rainforest are burned down to create land for farming, leaving fewer trees to absorb carbon dioxide. Burning releases the carbon that was stored in the trees too.

Around 200,000 acres of rainforest are destroyed every day.

25

Deserts

All deserts have a dry climate. Most of the time, they are too dry for plants to grow.

Many people think deserts are constantly hot, but in fact they become very cold at night. In the daytime, the heat is intense because there are few clouds to block the Sun's rays. Temperatures can reach 52°C. At night, the heat escapes quickly from the land, and temperatures dip below 0°C.

INTERNET LINKS

For links to the following websites, go to **www.usborne-quicklinks.com** and enter the keywords "science weather".

Website 1 Lots of information about deserts, including the Namib in Africa.

Website 2 Find out about the prairie, a type of dry grassland in North America.

Websites 3 and 4 Explore temperate and boreal forests, with pictures.

Tropical grasslands

Tropical grasslands, also called savanna, are flat, open plains. They have two seasons: rainy and dry.

The dry season lasts for about six months. Drought is followed by fierce storms.

After the storms have passed, the rainy season begins. Heavy rain falls and lush grasses begin to grow.

Dry grasslands

Dry grasslands are mostly found in central America and central Asia. They have a long, dry season, and a shorter wet season.

Dry grassland names

You might hear different names for dry grasslands around the world.

Place	Grassland
Africa	Veld
USA	Prairie
Russia	Steppes
Argentina	Pampas

Pampas grass

Farming and climate

People try to make farmland out of some grassland areas. The wet season is unreliable, though, so crops often fail. After a while, the farmers give up and move on, and the land becomes desert. The hot ground dries the air and even less rain falls, creating a drier climate.

Overfarming turns grassland to desert.

This desert in Namibia, Africa, has a permanently dry climate. But others have rainfall at certain times of the year, and they come alive with plants.

Temperate regions

The area between the Arctic and Antarctic Circles, which lie at about 66° north and south, and the Tropics is called the temperate zone. Temperatures here are neither very hot nor very cold, and there are four seasons – spring, summer, autumn and winter. Temperate climates can be warm, cool or cold.

Many trees in cool temperate zones shed their leaves each year. They grow back when the weather warms up in spring.

Warm temperate

Warm temperate or Mediterranean climates exist between about 20° and 40° north and south of the equator. They have hot, dry summers and warm, wet winters. In summer, dry, tropical winds bring high temperatures and little rainfall. During the winter, the winds change direction and rain falls.

Cool temperate

Areas that lie between 40° and 60° north and south have a cool temperate climate. Here, cold air from the poles clashes with warm air from the Tropics, leading to cloudy or changeable weather all year long. Most of Europe has a cool temperate climate.

Cold temperate

Cold temperate climates exist in the boreal forest or taiga, a wide area of frozen land and coniferous trees in Northern America, Europe and Asia. The long, harsh winters last up to six months, with average temperatures below freezing. Summers are short, with only 50 to 100 frost-free days.

This is a boreal forest in Alaska. Fewer things live in boreal forests than in any other type of forest. It's just too cold there.

27

Polar regions

The areas around the North and South Poles have frozen polar climates. Polar regions are very windy. The cold, heavy air above them sinks and presses down hard, pushing away all the air that was underneath. This blows outwards, and the strong winds make the area feel even colder than it really is. This effect is called wind chill.

Sinking cold air creates area of high pressure.

Wind blows outwards from high-pressure area.

Winds

See for yourself

You can feel the effects of wind chill for yourself on a cold, blustery day. It feels colder when you stand in the wind than when you are sheltered from it.

This is because the heat of your body warms a thin layer of air around you. Cold wind cools this air down, and you feel a chill.

Thick clothes, such as gloves, help to trap a layer of heat around your body.

Although polar regions are constantly sunny for almost six months every year, average temperatures never rise above freezing. This is because the Sun's rays are thinly spread, and most of them are reflected away from Earth by the ice and snow.

It takes a special kind of animal to live in the frozen landscape of Antarctica. Penguins have thick skin to protect them from the bitter cold.

Global warming is melting the vast sheets of ice which polar bears and their prey live on. The bears are starving because there's less prey for them to catch.

In the tundra, there's very little food available in winter. Many animals, including this snowy owl, move to warmer areas in the coldest months.

The frozen tundra

Tundra climates exist just north of the Arctic Circle. They have short, chilly summers, but long, bitterly cold winters. In fact, the weather warms up a little for only two to four months each year. Even the ground is frozen, forming a mixture of soil and ice called permafrost.

In spite of the cold, very little snow falls in the tundra, but strong winter winds blow the powdery snow around, creating chilling blizzards. Trees and other large plants can't grow in the tundra because of the high winds and frozen soil.

Up a mountain

In mountain regions, temperatures drop the higher you go. Trees can't grow beyond a certain level on a mountain because it's too cold and windy. This level is called the treeline. The cold area that exists above the treeline is called an alpine biome.

Places with alpine climates are often bare or covered with snow and blasted by freezing winds. It feels a lot like the tundra.

This Swiss valley has a pleasant, sunny climate, but if you were standing on the mountain top, you'd be freezing. You can see how trees stop growing at the height where the climate changes.

INTERNET LINKS

For links to the following websites, go to **www.usborne-quicklinks.com** and enter the keywords "science weather".

Websites 1 and 2 In-depth information about climate, plants and animals in tundra and alpine biomes, with lots of images.

Website 3 Fascinating facts about Antarctica, with an interactive map.

Website 4 Get Arctic weather reports and compare temperatures on a world map.

Local climates

Sometimes, a small area has a different climate from the places around it. This is called a local climate. Many things help to create local climates, including the shape of the land and how far it is from the sea.

Mountain climates

Mountainous areas can contain lots of different local climates. For instance, mountains are colder at the top than at the bottom. A mountain with a frozen peak may even have a tropical climate at its foot.

A mountain range can have a rainy climate on one side but a dry one on the other. This is because the range interrupts air flow over the land, affecting the climate.

Warm valley air rises. This leads to rain and thunderstorms.

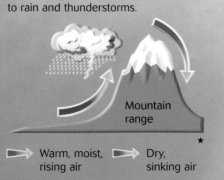

Mountain range

Warm, moist, rising air

Dry, sinking air

As warm, moist air rises up the side of the mountain range, it eventually dries out. It sinks on the other side, creating a dry climate there.

Forest climates

Forests often have a cooler, calmer local climate than surrounding areas, because leaves scatter sunlight and trees block winds. This effect changes according to the kind of trees in the forest, and the time of year.

In this picture of the Himalayas, you can see that conditions are clear and cold around the peaks. It becomes foggy and damp in the valleys below.

Coastal climates

Around the coast, the air is constantly warming up and cooling down. It rises and sinks in cycles, creating a mild, breezy climate, called a coastal or maritime climate.

During the day

Land warms up more quickly than sea. This heats the air above it, which rises.

Over the sea, the warm air cools, then sinks again.

Land

Sea

An onshore breeze blows from the sea to the land. ★

During the night

Sea cools more slowly than land. Warm air rises from it.

Air cools over land, and sinks.

An offshore breeze blows from the land to the sea. ★

Microclimates

Microclimates are even smaller than local climates. They are created by conditions in a very small area, such as the shady side of a building or the inside of a greenhouse.

In cold countries, you can grow tomatoes in a warm microclimate, such as in a greenhouse or a sunny, sheltered spot in a garden.

City climates

A city has a warmer local climate than the areas around it. This is because cities are mostly made of concrete, which absorbs lots of heat from the Sun during the day. At night, the concrete loses heat slowly, so nights often feel warmer in a city than in the countryside. City climates are also called urban climates.

INTERNET LINKS

For links to the following websites, go to **www.usborne-quicklinks.com** and enter the keywords "science weather".

Website 1 More information about microclimates, with images.

Website 2 Mountains of the world.

Warm city air creates a blurry heat haze around the buildings.

Water and oceans

We often take water for granted, but it's one of Earth's most important features. Without water, there would be no plants or animals, and no weather. The way that heat and water move between the oceans and the air creates the world's weather and climates.

Fresh and salt water

Earth has two types of water: salty sea water and fresh water. About 97% of all water is sea water. Most fresh water is frozen in ice sheets. In fact, only 1% of the water on our planet is fresh water that we can use.

The oceans are full of life. Lionfish like this live in the warm waters of the Pacific Ocean. There may be up to five million types of sea creatures still undiscovered.

The water cycle

Water is constantly moving between the oceans, air, land and living things. This is called the water cycle.

Water evaporates from the sea, that is, it turns into water vapour. As it evaporates, water loses its saltiness, becoming fresh.

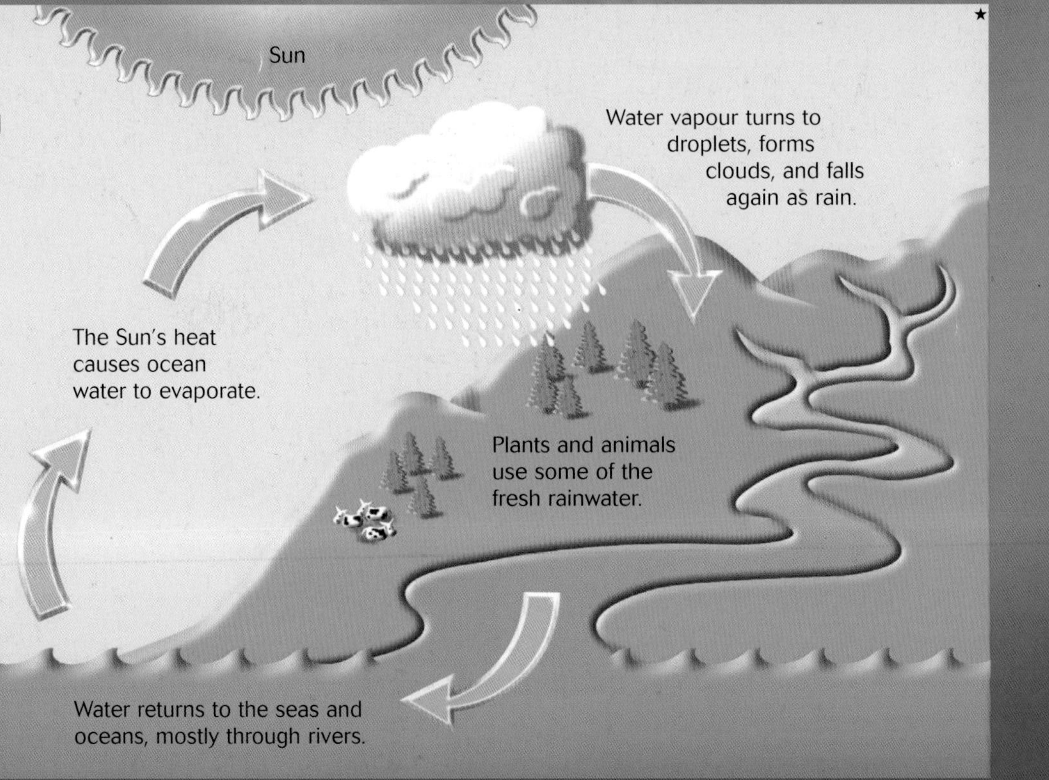

This diagram shows how water moves from the oceans to the air, to land, and back to the oceans.

Sun

Water vapour turns to droplets, forms clouds, and falls again as rain.

The Sun's heat causes ocean water to evaporate.

Plants and animals use some of the fresh rainwater.

Water returns to the seas and oceans, mostly through rivers.

Ocean currents

Sea water doesn't stay still. It travels around the world in huge bands called currents, which flow between warm and cold areas. Earth's main currents flow between the equator and the North and South Poles.

Below, you can see what happens at each of the numbered points on the globe on the right.

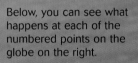

North Pole

3

2

Equator

1

2

3

South Pole

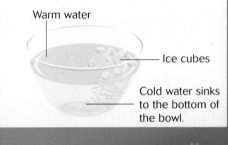
1   Equator   **2**   **3**   Pole

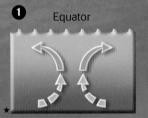

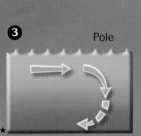

At the equator, cold water from the poles arrives in the ocean depths. It warms up and rises to the surface.

The warm water becomes a surface current. It flows towards the poles, often pushed along by winds.

At each pole, the warm surface water cools and sinks. It begins to flow back towards the equator.

Moving warmth around the world

A mass of water can store more heat from the Sun than the same mass of air can. The oceans store lots of heat, which is moved around the globe in warm currents.

Warm ocean currents heat the air above them too. If the air moves over land, it warms the local climate. Cold water currents have the opposite effect, cooling the air and the local climate. Without the winds and currents, many places on Earth would become much hotter or colder.

This coral reef grows in waters warmed by currents flowing from the equator towards the coast of Australia.

Earth's major currents

All currents affect climate. The main current in the southern hemisphere is the Antarctic Circumpolar Current (ACC), which flows around Antarctica. It acts as a "conveyor belt", mixing the waters of the Pacific, Atlantic and Indian Oceans and helping to control their climates.

The main current in the Atlantic Ocean is the Gulf Stream, a warm current which flows from the Gulf of Mexico to northwest Europe. Along with the North Atlantic Drift current, it makes up the North Atlantic Conveyor, a system which controls north European climate.

Locations of major currents

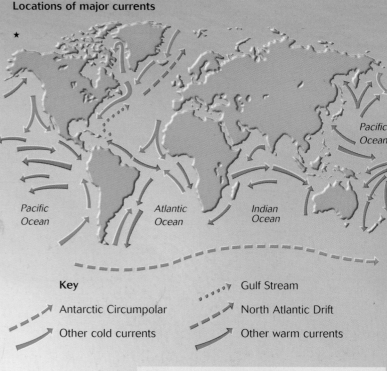

Pacific Ocean

Pacific Ocean

Atlantic Ocean

Indian Ocean

Key

Antarctic Circumpolar

Other cold currents

Gulf Stream

North Atlantic Drift

Other warm currents

Keeping a continent warm

Northwest Europe is warmer than Canada, although both areas are similar distances from the North Pole. The difference is that northwest Europe has the warming effects of the North Atlantic Conveyor.

These Inuit people live in northern Canada, where there's snow on the ground all year round.

Current threat

The North Atlantic Conveyor brings warm water to the North Atlantic. But the system slows down if the water is too warm, or not dense and salty enough, to sink when it arrives. Global warming can cause a slowdown: by heating the water, and by diluting salty water with too much fresh water from melting ice.

Global warming may be causing glaciers to melt, adding fresh water to the sea and slowing down the North Atlantic Conveyor.

INTERNET LINKS

For links to the following websites, go to **www.usborne-quicklinks.com** and enter the keywords "science weather".

Website 1 An interactive explanation of the causes of ocean currents.

Website 2 Watch a great animation that tells you more about the water cycle, and take a quiz too.

Website 3 Read about the effects of the El Niño in 1997-98, with videos.

El Niño

El Niño is the name given to a pattern of extreme weather in the tropical Pacific, which happens at least twice every ten years. It's called *El Niño*, meaning "the Christ-child", because it tends to begin around Christmas.

El Niño occurs when the usual flow of winds and currents weakens or even reverses.

Extreme weather is a threat to sea animals like this harbour seal. It makes it harder for mothers to find food, so their cubs starve.

A normal year

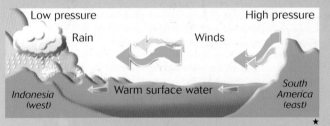

Normally, pressure is high in the east and low in the west. This creates winds which move warm surface water west towards Indonesia.

An El Niño year

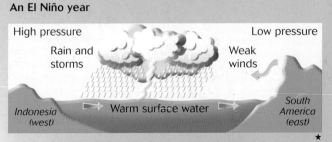

During an El Niño year, pressure reverses, and the winds weaken or even reverse. The warm water starts to move east, bringing heavy rain and storms.

La Niña

An El Niño event is often followed by a period of unusual cooling in the eastern tropical Pacific. This is called La Niña, and its effects are opposite to El Niño.

During La Niña, extra-strong winds push more warm surface water than usual westwards towards Indonesia. Around the Americas, cold deep-sea water rises to fill the space.

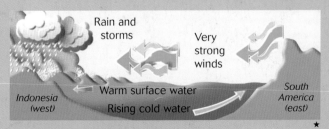

Less evaporation takes place from cold water so, during La Niña, rainfall and storms happen less often in the east, but more frequently over the warm waters gathering in the western Pacific.

Clouds

Clouds are made of millions of tiny water droplets or ice crystals, created when air rises and cools. Although they often bring rain, snow or hail, not all clouds lead to wet weather.

Water in the air

Water is constantly entering the atmosphere as part of the water cycle. When oceans, rivers and lakes become warm enough, water evaporates from them and rises from their surface as water vapour.

When water vapour cools down, it turns back into drops of liquid water. This is condensation – the opposite of evaporation. Water vapour condenses at a temperature called the dew point. When water condenses in the air, it makes clouds.

As water vapour rises, it cools.

Water rises into the air as water vapour.

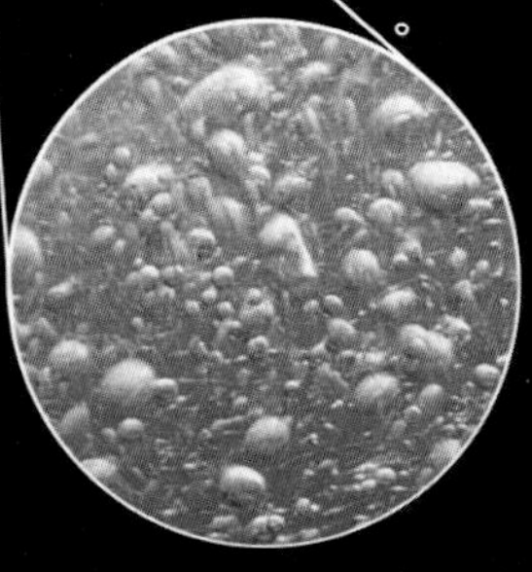

Cool water vapour turns to water drops, forming clouds.

Some clouds are made of ice crystals such as these, shown highly magnified. The blue areas are unfrozen water and other particles from the air.

Dark, puffy clouds like this often bring rain.

Making clouds

If a warm patch of land heats the air touching it, a "bubble" of warm air can form. As this rises through the atmosphere, the water vapour in it cools and turns into cloud.

Clouds also form when air rises to travel over an area of high land.

Moving air cools as it's forced to rise over higher land.

Hill clouds form as water vapour in the air condenses.

Cloud cover

Cloud cover is measured in oktas – a scale of 0 to 8. A sky with 8 oktas of cloud is completely covered. A sky with 0 oktas of cloud is clear.

0 oktas – clear sky

4 oktas – half covered

8 oktas – total cover

INTERNET LINKS

For links to the following websites, go to www.usborne-quicklinks.com and enter the keywords "science weather".

Website 1 An interactive guide to clouds, with games and an experiment.

Website 2 Play a cloud game, and find out more about what clouds do.

Layers of cloud

In places which give out lots of heat, such as cities, temperature increases with height. This effect, called a temperature inversion, makes clouds form in layers.

Warmer air

A rising bubble of air, already cooled below its dew point, meets a layer of warmer air.

Warmer air

The cooler air is trapped under the warmer air and spreads out.

Warmer air

More air bubbles rise. More layers of cloud form.

Feedback systems

In a feedback system, the effect of an action "feeds back" and affects the cause. In the example below, a strengthened greenhouse effect creates more clouds, which in turn strengthen the greenhouse effect more, and so heat up the Earth.

Greenhouse effect increases.

This makes oceans warm up, and more clouds form.

Clouds trap heat around Earth, increasing greenhouse effect more.

Trap or reflect?

Some clouds also reflect sunlight back into space, so they can help lower global temperatures too. Whether clouds trap or reflect heat depends on their height, type and thickness.

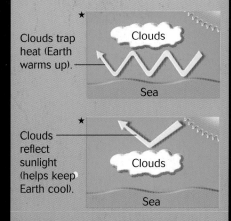

Clouds trap heat (Earth warms up).

Clouds

Sea

Clouds reflect sunlight (helps keep Earth cool).

Clouds

Sea

Because of this dual effect, it's not easy to predict how clouds will affect climate.

Cloud types

Clouds come in many shapes and sizes. To help identify them, they can be divided into four groups, based on the height at which they form.

High-level clouds form high in the troposphere, usually above 6,000m. They are wispy and made of ice crystals.

Medium-level clouds form halfway up the troposphere. They are white sheets or small clouds of ice and water drops.

Low-level clouds exist low down in the troposphere, below 2,000m. They are made up mainly of water drops.

Some clouds grow upwards, so they cross all levels. They have a puffy shape, and may contain water or ice crystals.

These strangely shaped clouds are called lenticular clouds. They sometimes form over mountains where air flows fast over uneven land. Some have even been mistaken for UFOs.

Cloud patterns

Each of the four cloud types can form lots of different shapes. The cloud shapes are named according to what they look like: for example, "cirro" means wispy, "strato" means layered, and "cumulo" means puffy. Clouds which bring very wet weather have "nimbus" in their names.

INTERNET LINKS

For links to the following websites, go to **www.usborne-quicklinks.com** and enter the keywords "science weather".

Website 1 An interactive look at different types of clouds.

Website 2 A gallery of cloud images.

▲ Cirrocumulus clouds are individual clouds of ice crystals, which often appear in a row.

▲ Cirrostratus clouds form a thin, almost transparent layer over the whole sky. They can often be seen before a storm.

▲ Cirrus clouds look feathery.

Cumulonimbus clouds have ▶ a billowing shape. They grow very tall, developing a flat top when they reach the limit of the troposphere. They bring heavy rain and often thunder.

Altocumulus clouds are made of small grey or white cumulus clouds of similar sizes. They often form in rows. ▼

Altostratus clouds usually make a grey sheet across the sky. ▼

Stratus clouds sit in a layer across the sky, often producing rain or drizzle. ▼

Stratocumulus clouds are a mixture of layered and puffy clouds. They are made from a sheet of rounded cumulus clouds which are almost joined together. ▼

Cumulus clouds are often seen on dry, sunny days. ▼

Nimbostratus clouds appear as a tallish, thick, grey layer, with an uneven base. They bring rain. ▼

13,000m

High level

6,000m

Medium level
Clouds at this level have "alto" in their names.

2,000m

Low level

500m

Water in the air

If the water droplets or ice crystals that make up clouds join together and become too heavy, they fall as rain, hail, snow or sleet. Together, all these kinds of wet weather are called precipitation.

Rain

Rain can be made either from water droplets in low-level clouds, or from melted ice crystals in higher clouds.

Very light rain is called drizzle. Sleet is a mixture of rain and snow, made when rain falls through cold air and partly freezes, or snow falls through warm air and partly melts.

Rainbows

A rainbow appears when sunlight shines through water droplets in the air. White light is made of seven colours: red, orange, yellow, green, blue, indigo and violet. As it shines through water, the light is bent. Each of the colours bends by a different amount, so they separate and become visible.

Rain provides the clean, fresh water that almost every living thing on land needs to live.

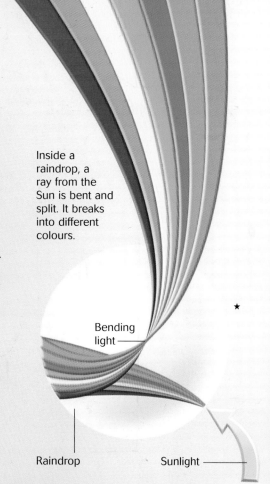

Inside a raindrop, a ray from the Sun is bent and split. It breaks into different colours.

Bending light

Raindrop

Sunlight

★

See for yourself

This is how you can create your own miniature rainbow.

1. Put a bowl of water in a very sunny place. Prop a small mirror up against the inside of the bowl.

2. Adjust the mirror so that a strong beam of sunlight falls on it. Move a piece of paper around in front of the bowl, until a reflected "rainbow" appears on it.

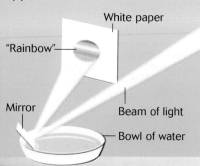

White paper

"Rainbow"

Mirror

Beam of light

Bowl of water

These snowflake images have been coloured using a computer, to show their structure. All snowflakes have six sides but no two are identical.

Hail

Hail is made of lumps of ice. It forms in cumulonimbus clouds called thunderheads.

Strong air currents carry water drops up into very cold parts of the atmosphere. There they freeze, and are tossed up and down, becoming coated with layer after layer of ice. They eventually fall as hailstones.

In this cross-section, you can see how a hailstone is built up of many layers.

Snow

Snow is made up of large ice crystals. These form when water drops join with much smaller ice crystals and then freeze. When it's very cold, snow falls in small, dry flakes. In warmer conditions, you'll see large, wet flakes instead.

A short, light snowfall is known as a flurry. Heavy snow in wind is called a blizzard.

INTERNET LINKS

For links to the following websites, go to **www.usborne-quicklinks.com** and enter the keywords "science weather".

Website 1 Watch an animated explanation of rainbows.

Website 2 Find out more about snow and see snowflake photos and movies.

A wetter future?

Some experts predict that in the future, global warming will bring more wet weather and heavier rainfall in many areas. This is a result of the water cycle becoming stronger. As global warming heats the oceans, more water evaporates, then condenses to form rainclouds.

This picture shows an unusually fierce downpour over part of Australia.

Mist, haze and smog

Mist is a thin type of fog, and haze and smog contain particles of dust or smoke as well as tiny water drops. Like fog, they all affect visibility – the distance you can see ahead.

Mist forms in a similar way to fog, but it's less dense. You can often see over the top of mist, because it stays closer to the ground.

Dew

On clear nights, the ground loses heat quickly and cools the air above it. Water vapour in the cooling air condenses, coating everything at ground level with tiny water droplets. This is dew.

These droplets of dew formed when water vapour condensed onto leaves.

Haze is a layer of fine dust and extremely tiny water drops. It forms in calm, dry conditions in cities and industrial areas.

Fog

Fog is like ground-level cloud. It usually forms when moist air travels over cold land or sea. The moist air cools down and the water vapour condenses near to the Earth.

Fog often appears during the evening or early morning. It's most common around the coast and in valleys, where it forms from sinking mountain air.

Fog tends to develop on clear, cool nights, when moisture escapes from warm ground which is quickly cooling down.

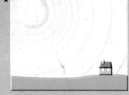

The longer and cooler the night, the more fog will form. It soon disappears on a clear morning if the Sun heats the ground.

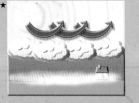

If cloud forms before morning, the sunlight is blocked. In this case, the ground doesn't warm up, so it might stay foggy all day.

Smog (**sm**oke and f**og**) is a form of extreme fog. It is made when smoke or gases from vehicle exhausts join with water droplets in the air.

Frost

Frost develops when it's cold enough for water vapour near the ground to condense onto objects as ice instead of dew. If the ground itself is chilled to below 0°C, ground frost forms. When the air above the ground is below 0°C, it's called an air frost. On this page, you can see three different kinds of frost.

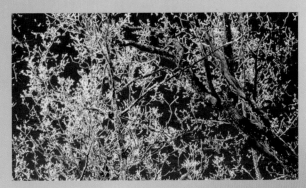

Rime is a spectacular, icy crust which forms in a wet wind in temperatures below zero. The air carries very cold water droplets, which freeze on any object they touch.

Hoar frost is made when water vapour turns directly into spiky ice crystals.

INTERNET LINKS

For links to the following websites, go to **www.usborne-quicklinks.com** and enter the keywords "science weather".

Website 1 Photographs and facts about how frost forms.

Website 2 Watery weather topics, including humidity and dew, with activities to try.

Fern frost forms on windows, when dew on the cold glass turns to ice. More moisture then freezes on top of it in layers, creating feathery patterns.

Weather and air

The temperature of air is affected by the surface it moves over. For instance, polar land and water chill the air above them. At the equator, warm water and land heat the air. All our weather is caused or affected by the temperature of the air and how it moves around.

Land, sea and air temperature

Land and sea affect air in different ways. The Sun only heats land to a shallow level, so it heats up and cools down quickly. The air above also heats and cools quickly.

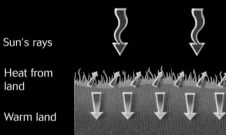

Sun's rays

Heat from land

Warm land

In the ocean, heat is moved away from the surface, so the water is warmed more deeply than land. The heat is released slowly into the air above.

Sun's rays

Heat from water

Warm water

Air that travels over sea is humid – that is, it contains moisture – because water vapour rises into it. Air that's moved over land is usually drier.

Air currents

Air that's warmed by the land or sea begins to rise. As it gets higher, it cools and sinks back to Earth. There, it may be heated again if the land's surface is still warmer than the air itself.

The circulation of warm and cold air is known as convection, and the moving streams of air it creates are convection currents. These control Earth's temperatures by moving heat between the equator and the poles.

A paraglider uses rising warm air to stay in the air for a long time.

How a convection current works

Rising current of warm air

Warm air cools and sinks.

Cooler air moves in to replace rising warm air.

INTERNET LINKS

For links to the following websites, go to **www.usborne-quicklinks.com** and enter the keywords "science weather".

Website 1 Find out what causes weather, with animations and quizzes.

Website 2 Ride a wave of air over a mountain in a virtual sailplane.

Low and high pressure

When warm air rises, it leaves behind an area of low pressure. As cool air sinks, it creates an area of high pressure below it.

Low-pressure areas, also known as depressions or cyclones, cause changeable weather. High-pressure areas, also called anticyclones, bring fine, warm weather in summer, and cold but calm weather in winter.

Paragliders gain height on currents of rising warm air called thermals. These form over patches of heated land.

Air pressure

Air presses on things in all directions, with a force called air pressure or atmospheric pressure. This is about $1kg/cm^2$ at sea level, but it changes with height. It's measured using a barometer. Air pressure plays an important part in creating different weather conditions, especially winds.

Low pressure   **High pressure**

As warm air rises from the surface, more air moves in from higher pressure areas.

As cooler air pushes down, the surface air moves away to lower pressure areas.

An inflated balloon has equal air pressure on the inside and outside.

Pressure

Pressure

If you stick a pin in a balloon, the pressurized air inside quickly escapes. The balloon bursts as the air flies out.

Winds

Differences in temperature and pressure cause air to move around, making winds blow on a huge scale. All winds blow from areas of high pressure to areas of low pressure.

Air movement

There are several major bands of high or low pressure around the Earth. The most important one is around the equator, where air is heated up more than anywhere else. As it rises, the warm air leaves behind a low-pressure band called the Intertropical Convergence Zone (ITCZ). This affects all other global air movements.

General movement of air around the world

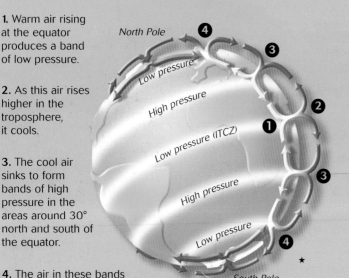

1. Warm air rising at the equator produces a band of low pressure.

2. As this air rises higher in the troposphere, it cools.

3. The cool air sinks to form bands of high pressure in the areas around 30° north and south of the equator.

4. The air in these bands meets with colder, denser air moving away from the poles. It is forced to rise again, creating bands of low pressure at around 60° north and south. These are areas of unsettled weather.

North Pole

Low pressure

High pressure

Low pressure (ITCZ)

High pressure

Low pressure

South Pole

Local winds are always blowing in sunny coastal areas. Good conditions for sailing and surfing are more reliable here than on an inland lake.

Local and global winds

The movement of air around the globe creates winds which blow in the same direction and across the same region most of the time. These are called prevailing winds.

As well as prevailing winds, there are also local winds, which exist on a much smaller scale. These blow between smaller areas of high and low pressure which move around over the land and sea.

INTERNET LINKS

For links to the following websites, go to **www.usborne-quicklinks.com** and enter the keywords "science weather".

Website 1 An interactive guide to air movements and jet streams, with animations and fascinating facts.

Website 2 Try online activities that demonstrate how the Coriolis effect works.

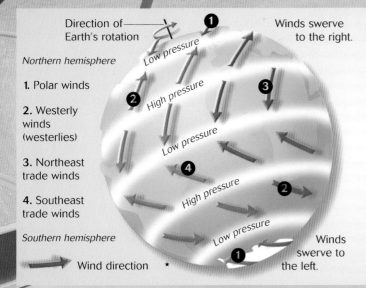

Swerving winds

If Earth stood still, its prevailing winds would blow directly from north to south or south to north. But Earth spins like a top, causing the winds to swerve sideways. This is called the Coriolis effect. Winds in Earth's northern half swerve to the right of the direction they would otherwise take. Winds in the southern half swerve to the left.

The diagram on the left shows the Coriolis effect and the world's prevailing winds.

Direction of Earth's rotation

Winds swerve to the right.

Northern hemisphere

1. Polar winds

2. Westerly winds (westerlies)

3. Northeast trade winds

4. Southeast trade winds

Southern hemisphere

→ Wind direction

Low pressure

High pressure

Low pressure

High pressure

Low pressure

Winds swerve to the left.

Jet streams

Differences in air pressure occur in the troposphere where cold air and warm air meet. This produces strong winds called jet streams at the top of the troposphere.

The strongest jet streams are found at roughly 60° north and 60° south, where high-level polar air meets warmer air. There are weaker jet streams at 30° north and 30° south.

Jet streams occur about 10km above the Earth, near the tropopause.

Polar front jet stream. Blows at an average speed of 50-110kph.

Subtropical jet stream. Less powerful than polar front jet stream.

60°N
30°N
Equator
0°
30°S
60°S

Jet stream paths

Jet streams disrupt air on either side of their paths, which affects air movement nearer to the Earth. They drag areas of high and low pressure along with them.

The air movements that jet streams produce are responsible for the surface winds we feel from day to day in many parts of the world.

How jet streams affect weather

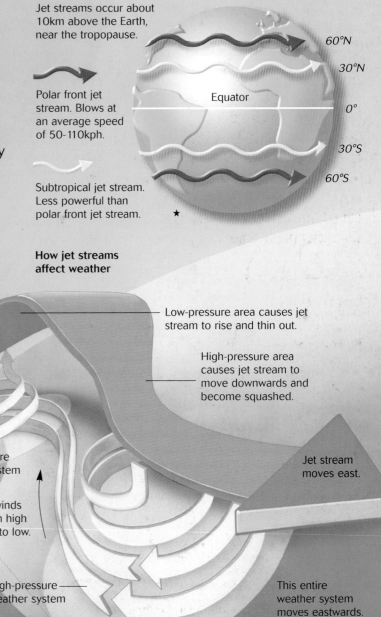

Low-pressure area causes jet stream to rise and thin out.

High-pressure area causes jet stream to move downwards and become squashed.

Jet stream moves east.

Low-pressure weather system

60°N

West

East

Surface winds blow from high pressure to low.

High-pressure weather system

This entire weather system moves eastwards.

47

Air masses and fronts

An air mass is a huge "lump" of air with the same temperature and humidity throughout. Air masses move around, changing the weather as they go.

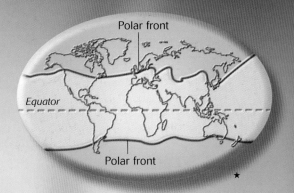

Polar front

Equator

Polar front

This map shows the general areas of warm and cold air around the Earth. The boundaries between them are called polar fronts.

How air masses form

Air masses form when air rests over land or sea with an even temperature or humidity. The air takes on these features itself. Air masses are named after where they form: maritime (m) air masses form over sea, and continental (c) air masses form over land.

The map below shows examples of air masses, and how they are named.

Moving air masses

Air masses are affected by the surfaces they pass over, and the speed at which they travel. Slow-moving air has more time to be affected by the surfaces it moves over.

If an air mass is warmer, colder, drier or more humid than the air in the area it moves into, it will make the weather change.

Fronts

A front is a narrow boundary between two air masses with different temperatures and humidity levels.

Fronts are on the move all the time, as one air mass replaces another. On a worldwide scale, though, the boundaries between cold polar air and warm tropical air stay in roughly the same place. These major boundaries are known as polar fronts.

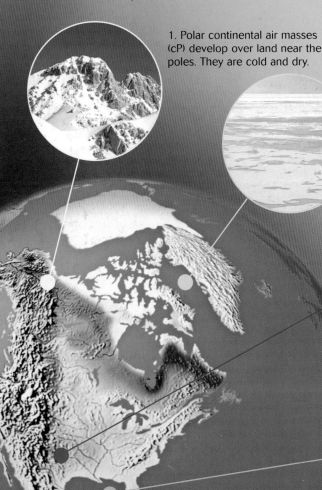

1. Polar continental air masses (cP) develop over land near the poles. They are cold and dry.

2. Polar maritime (mP) air masses form over the sea near the poles. They are cold and moist.

3. Tropical continental (cT) air masses develop over hot, dry land. They are warm and dry.

4. Tropical maritime (mT) air masses develop over warm seas. They are warm and moist.

This diagram shows how a depression is drawn on a weather map. The white lines, called isobars, join together places with the same air pressure.

◁ Cold front
◁ Warm front
◯ Isobars
↘ Wind direction

Low-pressure centre

Depressions

Fresh masses of cold and warm air are constantly produced at the poles and the Tropics. But the two air types do not easily mix. Where they meet, the warmer air tends to rise, so a low-pressure area called a depression occurs.

Depressions are also known as cyclones or lows. They bring rain, storms and changing temperatures.

Winds blow almost exactly parallel to the isobars.

The closer together the isobars are, the stronger the winds.

INTERNET LINKS

For links to the following websites, go to **www.usborne-quicklinks.com** and enter the keywords "science weather".

Website 1 Good, basic information about fronts and air masses.

Website 2 Air mass satellite movies.

Both fronts are advancing in the same direction.

Joining fronts

Cold fronts move faster than warm fronts. Sometimes, a cold front catches up with a warm front and they join together. This creates an occluded front, or occlusion.

Warm and cold fronts

There are two types of fronts in a depression. A warm front is made by warm air pushing into a cold air mass and rising over it. A cold front bulges into a warm air mass from below.

How a warm front forms

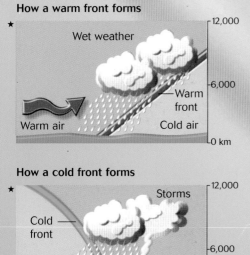
Wet weather
Warm front
Warm air
Cold air
12,000
6,000
0 km

How a cold front forms

Storms
Cold front
Cold air
Warm air
12,000
6,000
0 km

Moving depressions

Most depressions travel from west to east. In the northern hemisphere, if you stand with your back to the wind, air pressure will be lower on your left. In the southern hemisphere, this is reversed.

If you stood in one place while a depression passed overhead, you would see the weather change many times.

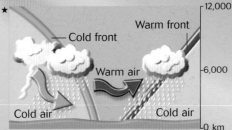
Cold front
Warm front
Warm air
Cold air
Cold air
12,000
6,000
0 km

As a warm front passes over, it rains. Then, the warm air mass brings calmer weather. Finally, the cold front brings cold weather, heavy rain and storms.

How an occlusion forms

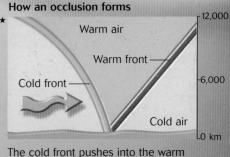
Warm air
Warm front
Cold front
Cold air
12,000
6,000
0 km

The cold front pushes into the warm air mass because it is moving faster.

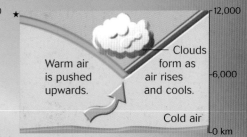
Clouds form as air rises and cools.
Warm air is pushed upwards.
Cold air
12,000
6,000
0 km

When the cold front reaches the warm front, the warm air mass rises and an occlusion forms, bringing storms. After an occlusion, the depression dies out.

49

Extreme weather

Sometimes, the weather turns wild. Fierce thunderstorms can form, or even hurricanes and tornadoes which leave behind a trail of destruction. Some of the most extreme storms are powerful enough to rip up a whole town.

Tornadoes

Tornadoes are violent, spinning, funnel-shaped winds that suck up objects from all around. In some places, they are known as twisters. A tornado that forms over water is called a waterspout.

A tornado forms in humid conditions, when warm and cool air meet in a huge, whirling storm cloud called a supercell. The air forms a spinning tube of wind which reaches down to the ground. Tornadoes bring heavy rain, thunder and lightning.

See for yourself

Next time a storm begins, find out how far away it is by counting the seconds between seeing a flash of lightning and hearing thunder. Each second indicates one mile. You hear the sound of thunder later because sound travels more slowly than light.

INTERNET LINKS

For a link to the following website, go to **www.usborne-quicklinks.com** and enter the keywords "science weather".

Websites 1 and 2 Build a tornado and see satellite images of extreme weather.

Tornadoes like this are mainly seen in North America. Winds can whirl at speeds of 120 to 510kph.

Hurricanes

In tropical areas, hurricanes (also called tropical cyclones or typhoons) may form over the ocean. They are huge, violent storms, often many hundreds of kilometres across, which bring torrential rain and heavy winds.

How a hurricane forms ★

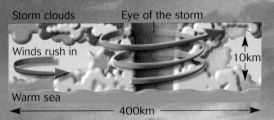

Storm clouds Eye of the storm

Winds rush in 10km

Warm sea

← 400km →

Moist, warm air rises over the sea. Winds from all around rush in and spiral up to replace it. As wind speeds increase, a fierce storm begins. It has a calm area called the eye in its middle.

Warm air from the Gulf of Mexico helped to create this hurricane in Texas.

Thunderstorms

Thunderstorms begin when very warm, moist air rises into cold air. Water droplets and ice crystals in clouds rub together, creating positive and negative electrical charges. This creates a build-up of electrical energy in the clouds.

Thunder is created by very hot lightning heating the air super-quickly. The air expands with a loud "boom".

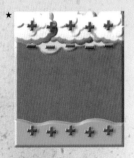

Negative charges gather in the base of the cloud. They attract positive charges in the ground, which gather below.

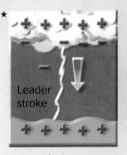

Leader stroke

Energy builds up in the cloud until it's pulled to the ground. This leader stroke creates a path between the cloud and the ground.

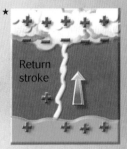

Return stroke

The main stroke, called the return stroke, follows the path back up. This is what we see as lightning and hear as thunder.

Floods

If lots of rain falls and there is too much to drain away, a flood will begin. Flooding can be caused by other things too, such as rising sea levels or waves swamping land during a storm. Snow on mountains can also cause floods lower down, when it melts rapidly into rivers, making them overflow.

INTERNET LINKS

For links to the following websites, go to **www.usborne-quicklinks.com** and enter the keywords "science weather".

Website 1 See how extreme weather affects people.

Websites 2 and 3 See the power of droughts and floods and the chaos they bring, with images, movies and more.

Droughts

A drought is a period of time when an area receives much less than its normal amount of rainfall. This may happen when an area of high pressure stops rain-bearing, low-pressure air from passing over. Droughts also occur when land is cleared of plants in places that are already very dry. Plants give out water vapour, so they help to create moist weather.

This dry, cracked earth in Thailand is part of a cleared forest. Without plants to give out water vapour, the air becomes drier. With little rain, the land dries up and drought begins.

Monsoons

In tropical areas, such as India and southeast Asia, winds blow in roughly opposite directions in different seasons. These monsoon winds can bring heavy rain or devastating drought.

Summer monsoon winds start to blow when the Sun is almost directly overhead. The land is heated intensely, warming the air above it. Moist air from over the sea rushes in to replace the rising warm air, and heavy rains fall. In winter, the Sun is no longer overhead and the land cools. Dry winds blow from the land out to sea.

Summer (rainy season)

A low-pressure area forms over the land as warm air rises. It is replaced by moist sea air, causing heavy rain.

Winter (dry season)

The land is now cooler than the water, so winds blow out to sea. They are dried out by the land they pass over.

This Vietnamese family's home is surrounded by flood water. Central Vietnam is regularly hit by severe flooding in the monsoon season.

Studying weather

Studying and predicting weather is called meteorology. Knowing what weather is on the way is important to many people: for instance, farmers will not plant crops if they are expecting frost or heavy rain.

Weather instruments

In the past, people used signs in nature, such as the colour of the sky at certain times of day, to predict the weather. Today, we use many different instruments to tell us about weather conditions. The main kinds are described on the next few pages.

A weathervane shows wind direction. It has a moving arrow which points in the direction the wind is coming from, and a fixed base marked with the compass points.

Anemometers measure wind speed. The cups catch the wind and make the central shaft spin. The number of times it rotates every minute gives the wind speed.

Anemometer

Cups to catch the wind

The number of rotations is recorded by a small counter at the base of the shaft.

Shaft

This is a French weathervane, so it has the initials for the French points of the compass – *Nord, Est, Sud* and *Ouest*.

See for yourself

Pine cones react to humidity, so you can use them to predict wet weather. On a dry day, their scales open out. If the air becomes moist, they close up.

INTERNET LINKS

For links to the following websites, go to **www.usborne-quicklinks.com** and enter the keywords "science weather".

Website 1 Find out more about instruments with a virtual tour of a weather office.

Website 2 Find out how to make many different kinds of weather instruments.

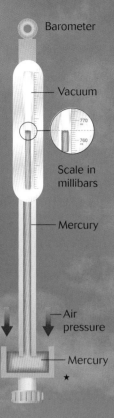

Barometer

Vacuum

770

760

Scale in millibars

Mercury

Air pressure

Mercury

★

Barometers measure air pressure. The barometer shown on the left is the traditional kind, with mercury inside. As air pressure increases, air pushes down on the mercury, forcing it up the tube.

Air pressure is measured in units called millibars. The average pressure at sea level is 1013.25 millibars.

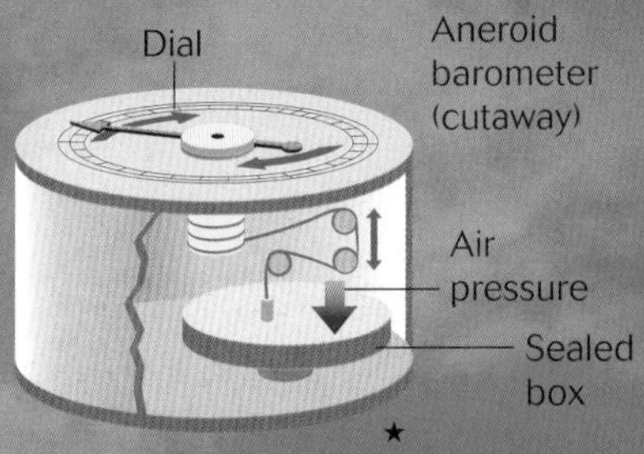

Change

RAIN

30

1020

1000

29

980

960

940

28

S

Stormy

Dial

Aneroid barometer (cutaway)

Air pressure

Sealed box

★

An aneroid barometer has a needle and dial above a sealed metal box. When air presses down on the metal box, a series of springs and levers move the needle.

Aneroid barometers like this are labelled with actual conditions, as well as millibar measurements. They are easier for non-meteorologists to use.

Hygrometer

Thermometers measure air temperature. As this changes, the liquid inside the tube expands or contracts. Expanding liquid moves up the tube and contracting liquid moves down.

Maximum thermometers measure the highest temperature reached. They contain mercury.

Mercury

★

0 10 20 30 40 50 60 70 80 90 100°C

Minimum thermometers measure the lowest temperature reached. They contain alcohol.

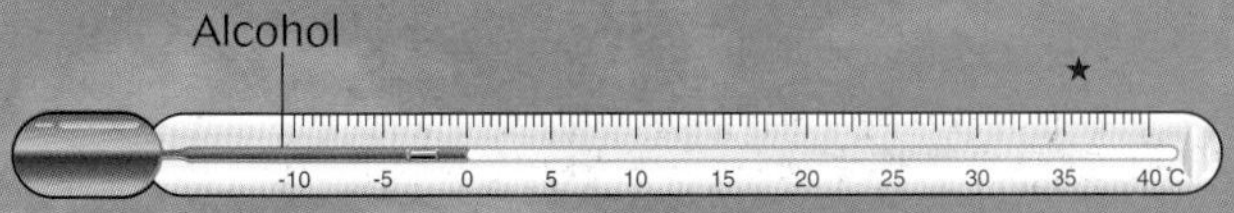

Alcohol

★

-10 -5 0 5 10 15 20 25 30 35 40°C

Hygrometers measure humidity. The most common type has two thermometers: one is dry and one is kept wet with a damp cloth. Water evaporating from the cloth cools the wet bulb, so the temperature shown is lower than the other thermometer.

If the air is dry, more vapour evaporates from the cloth. The difference between the thermometer readings becomes greater, which means humidity is low.

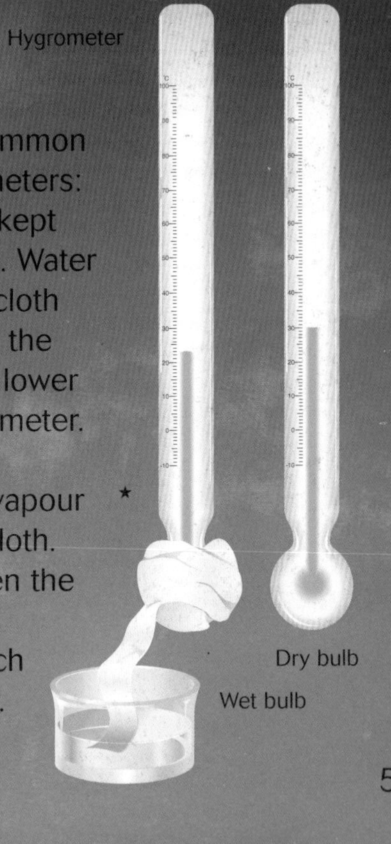

★

Dry bulb

Wet bulb

Weather forecasting

In order to predict the weather, experts constantly gather information about weather conditions everywhere. This information, called instrumental data, is collected using specialized equipment.

INTERNET LINKS

For links to the following websites, go to **www.usborne-quicklinks.com** and enter the keywords "science weather".

Website 1 Find out what happens at the Mount Washington Observatory.

Websites 2 and 3 Weather forecasting pages from the UK's Met Office.

Websites 4 and 5 Weather games.

Website 6 Weather forecasts around the world.

Weather stations

In a weather station, meteorologists use instruments to monitor wind speed, visibility, pressure, air temperature and rainfall. They also observe and record general weather conditions, such as fog and cloud type.

Most of Antarctica's weather stations are automatic, because the climate there is so harsh.

Automatic stations

Some places, such as polar regions and mountain tops, are just too inhospitable for people to live. Instead, automatic weather stations are set up, where computerized weather instruments, in the place of meteorologists, take readings at regular intervals.

Mount Washington, USA, experiences some of the most severe weather in the world. This weather station on its peak is occupied only in the summer, when conditions are milder.

Weather balloons

Instruments called radiosondes measure temperature, pressure and humidity at different heights above the ground. They are carried into the air by weather balloons.

The speed and direction of the rising balloon show the strength and direction of the wind. The radiosonde takes readings as it rises and transmits them to a weather station on the ground.

The balloon bursts when the air pressure surrounding it is less than the pressure inside. A small parachute drops the radiosonde gently to the ground.

This balloon will take about one hour to rise to 20km.

Parachute to control descent

Radiosondes are released from weather stations twice a day.

Weather satellites

Weather satellites orbit Earth, collecting information about global weather conditions. There are two main kinds of satellite: polar-orbiting and geostationary.

Polar-orbiting satellite

Geostationary satellite

→ Earth's rotation
➡ Path of satellite around Earth

Polar-orbiting satellites go around the Earth from pole to pole. They move slightly westward on each orbit, as Earth rotates.

Geostationary satellites stay in the same position over the Earth, by orbiting at the same speed as Earth rotates.

Data collection

Satellites use lots of equipment to gather information about Earth and its weather. Devices called radiometers show levels of light, heat or humidity around the Earth. They are used to create thermal images, like the one shown here.

This satellite image shows clouds (coloured pink) forming above warm seas. Maps like this are used to predict rain or droughts.

Radars

Weather radars identify areas of rain, snow or hail, and how heavily they are falling. A doppler radar, shown below, also detects the speed and direction of moving wet weather.

Antenna

Antenna sends out radio waves.

Receiving dish

If the waves hit stationary water drops, dust or ice crystals, they bounce back to a receiving dish at the same speed as they came out.

Clouds moving towards the antenna bounce the waves back to the receiver more quickly.

Clouds moving away from the antenna bounce the waves back more slowly.

Aircraft and ships

Some aircraft and ships carry equipment that records weather conditions along their routes. For instance, aircraft which fly at a height of between 10km and 13km provide information about high-level winds.

Creating weather forecasts

Before they can create a weather forecast, meteorologists gather data from weather stations, ships, balloons, satellites and radars. Then they interpret it, using computers. They publish their findings on a type of map called a synoptic chart, which is marked with symbols like the one below (there's a full list on page 87).

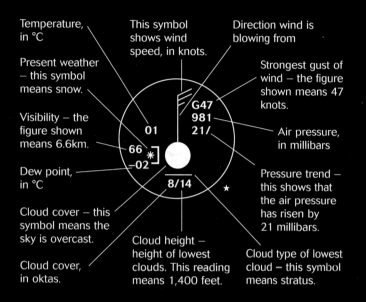

Temperature, in °C

This symbol shows wind speed, in knots.

Direction wind is blowing from

Present weather – this symbol means snow.

Strongest gust of wind – the figure shown means 47 knots.

Visibility – the figure shown means 6.6km.

G47
981
01   21/

Air pressure, in millibars

66 *
–02

Dew point, in °C

8/14

Pressure trend – this shows that the air pressure has risen by 21 millibars.

Cloud cover – this symbol means the sky is overcast.

Cloud cover, in oktas.

Cloud height – height of lowest clouds. This reading means 1,400 feet.

Cloud type of lowest cloud – this symbol means stratus.

Computer models

As well as synoptic charts, meteorologists produce computer models of the weather. Computers are programmed with readings of humidity, wind and air pressure at various points on an imaginary grid over the Earth. They can calculate new readings for any time in the future, based on the current set. But the further they go into the future, the less accurate they become.

Weather forecasts for areas within squares are calculated using the nearest grid point readings.

Weather conditions can vary with altitude, too. Each circle below shows a different height where a reading is taken.

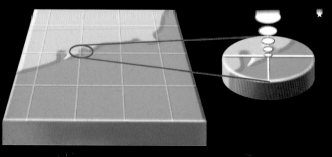

Weather maps

Weather forecasters use synoptic charts and computer models to compile weather maps. These simplified charts – the kind that you see on a TV weather report – show pressure, temperature, wind speed and other basic weather conditions, such as rain.

This weather map of Ireland shows cold, wintry conditions.

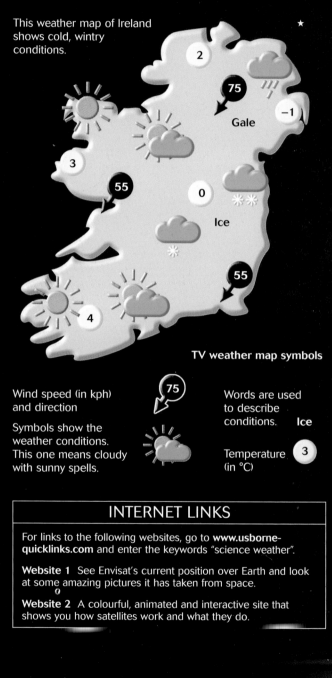

2
75
–1
Gale
3
55
0
Ice
55
4

TV weather map symbols

Wind speed (in kph) and direction

75

Words are used to describe conditions. **Ice**

Symbols show the weather conditions. This one means cloudy with sunny spells.

Temperature (in °C) 3

INTERNET LINKS

For links to the following websites, go to **www.usborne-quicklinks.com** and enter the keywords "science weather".

Website 1 See Envisat's current position over Earth and look at some amazing pictures it has taken from space.

Website 2 A colourful, animated and interactive site that shows you how satellites work and what they do.

Climate satellites

As well as helping us forecast weather from day to day, satellites can also be used to monitor climate and the environment.

In the USA, most of the data from climate satellites is collected by two organizations: the National Aeronautics and Space Administration (NASA) and the National Oceanic and Atmospheric Administration (NOAA). In Europe, data is gathered by the European Space Agency (ESA).

This is Envisat, Europe's most advanced environmental satellite. It circles the Earth every 100 minutes at a height of 800km.

New satellites

In August 2002, the ESA launched MSG1, the first in a new series of satellites called Meteosat Second Generation (MSG). MSG1 is in a geostationary orbit at 0° over west Africa. It monitors cloud, land and sea surface temperatures in part of the southern hemisphere, as well as water vapour, carbon dioxide and ozone in the atmosphere.

MSG1

Looking at the environment

Envisat (short for **envi**ronmental **sat**ellite) is an advanced polar-orbiting satellite, launched by the ESA in March 2002. Throughout its mission, it will provide measurements of atmosphere, ocean, land and ice. The data that Envisat sends back will allow scientists to study global warming, El Niño and the ozone layer in greater detail than ever.

Clues to past climates

Satellites and computers monitor climate all the time, but scientists can only decide if conditions are normal or not by working out what climates were like in the past. Clues in the natural world can help us to understand past climate patterns.

Studying the past

Scientists who study past climates are called paleoclimatologists. They try to find out what climates were like before people used special equipment to observe them. Since we can't measure past climates directly, we have to rely on evidence from rocks, trees, animal remains and the landscape. This kind of information is called proxy data.

Looking at fossils

The fossilized remains of ancient animals show us what conditions used to be like where they lived. For instance, remains of creatures that lived in cold seas can be found in places that are now very hot.

This 49-million-year-old python fossil was found in Germany. Today, the climate there is not hot enough for snakes of this kind to survive, but it must have been once.

Looking at land

Sometimes, features of the land's surface can tell us which places were once covered in ice.

Many U-shaped valleys, such as those found in England's Lake District, were made by glaciers when the climate was much cooler. Round lakes called kettle lakes, such as the ones in Cape Cod, USA, were left behind by chunks of glacial ice stuck in the ground. As the ice melted, it left a hole which filled with water.

This valley in Yosemite, California, was created many thousands of years ago by a glacier. Boulders were often picked up and carried by glaciers, then left in odd places when the glaciers melted.

Looking at rocks

Many rocks start out as layers of mud, sand and bits of stone called sediment, which build up over thousands or millions of years. By looking closely at these rocks today, scientists can find out things about the climate when the sediment was laid down.

Coal forms over millions of years in hot, steamy swamps. Today, though, it is mined in many places with a cool climate.

INTERNET LINKS

For links to the following websites, go to **www.usborne-quicklinks.com** and enter the keywords "science weather".

Website 1 An illustrated look at how and why we study past climates.

Website 2 Find out how coral reefs have been affected by climate change.

Website 3 How the real age of rocks is determined, with a slide show.

This red sandstone from southwest England formed over 200 million years ago. It has round sand grains, which tells us it started as a sand dune in a hot desert.

Sand grains, magnified

The types of fossilized sea creatures found in this limestone tell us about the temperature of sea water in the past.

Sedimentary rock

Sediment builds up on the ocean bed. Over millions of years, the huge pressure of water and more sediment on top turns it into sedimentary rock. Eventually, it is crumpled and lifted up by Earth's moving crust to form land, including mountains and other features.

Scientists can study both dry land and the ocean bed itself for clues to climate change over millions of years. Every layer of sediment tells its own unique story.

Each layer of this crumpled rock can tell us about the climate when it was deposited as sediment. For example, rocks which formed in warm climates often contain lots of fossils.

Written records

Records of past weather can help scientists to identify what range of weather conditions is "normal". Ancient inscriptions, travellers' journals, newspaper accounts and farmers' diaries are important sources of climate information. For example, experts used historical records of grape harvests to draw a graph of summer temperatures in Paris between 1370 and 1879.

These geese migrate when the weather is just right. Records of past migrations can show us if this time is changing from year to year.

Looking at trees

The rings in ancient tree trunks give clues to past weather patterns. One growth ring forms in each year of the tree's life. In years with a warm, wet growth season, the tree grows more and the rings are wide. During cold or dry years, there's not much growth, so the rings are narrow.

Wide growth rings show that the weather was probably warm and wet.

Narrow growth rings form in a cold or dry growing season.

Looking at pollen

Pollen grains can exist for many thousands of years without decaying. The types of pollen that have gathered over time in a particular place indicate what plants used to grow there.

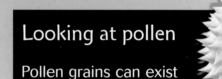

These are tiny pollen grains, magnified many times.

Climate change may be to blame for changes in an area's plant life, but it's hard to tell. Pollen can spread far from its parent plant, so the results of pollen analysis are not always accurate.

INTERNET LINKS

For links to the following websites, go to **www.usborne-quicklinks.com** and enter the keywords "science weather".

Website 1 Look inside the climatologist's toolbox to find many ways of examining climate.

Website 2 Find out more about tree rings and analyze a tree ring online.

Clues in ice

Scientists can compare current levels of greenhouse gases with past levels, by analysing ancient air bubbles trapped in glaciers and ice sheets. The deepest ice, in Antarctica, contains air and dust that's over 240,000 years old.

Scientists use a special drill to remove a long column of ice, called a core, from the dense layers. Air bubbles trapped in the core can tell us about levels of greenhouse gases when the ice formed.

Ice core

Air bubbles

★

Using the data

Scientists enter the proxy climate data described on these pages into computers to create climate models. These can help them understand past climates and also give clues to future climate change.

This scientist is cutting up part of an ice core sample. He wears protective clothing to avoid contaminating it.

63

Natural change

Although many people think that climate change is a man-made problem, most major changes to the climate system are natural ones. Many happen slowly, as a result of Earth changing and developing. Others are more extreme.

Early climate

Earth was originally a seething ball of hot, liquid rock. As it cooled down, a primitive atmosphere formed, and a climate which could support life began to develop. Since then, Earth has been through many periods of natural heating and cooling.

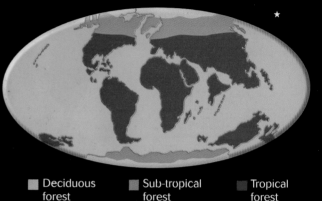

■ Deciduous forest   ■ Sub-tropical forest   ■ Tropical forest

Around 55 million years ago, rapid global warming made Earth a very warm place. Current temperate regions had tropical climates, and polar areas had temperate climates.

Sudden change

There is evidence of six occasions in Earth's past where many living things died out very quickly. For instance, at the end of the Permian period, half of all land species and 95% of sea species died out.

Events such as this, called mass extinctions, were probably caused by rapid climate change.

4,600 mya
Earth forms. No atmosphere, so no protection from Sun's rays.

2,000 mya
Oxygen begins to build up around planet. Major ice ages take place at 2,500 mya and 900-600 mya.

545-495 mya
Cambrian period
Continents exist in many pieces. Overall climate probably mild.

495-445 mya
Ordovician period
Mild at first, but by late Ordovician, most of southern hemisphere covered in ice.

445-415 mya
Silurian period
Warm conditions in northern hemisphere; lingering ice in southern hemisphere.

415-355 mya
Devonian period
Early: temperate and dry climates in many areas.
Middle: rainy equatorial areas; warm, shallow seas.
Late: One huge continent, Pangaea, begins to form.

355-290 mya
Carboniferous period
Tropical rainforests exist around equator, surrounded by deserts. Ice cap begins to extend from South Pole.

290-250 mya
Permian period
Early: most of southern hemisphere covered by advancing ice.
Late: rainforests replaced by desert. Ice in southern hemisphere retreats, but ice cap forms at North Pole. Pangaea has completely formed.

250-205 mya
Triassic period
Triassic period may be hottest in Earth's entire history. Poles have warm temperate climates. Pangaea starts to break up.

Climate change in the past

On this timeline, a cold average global temperature is shown in blue, and a hot average temperature is shown in orange.

Key

Mya = Millions of years ago

■ Mass extinction   ■ Beginning and end of eras

Change in space

Earth's climate can be affected over very long periods of time by changes to the Sun, and the way that Earth moves around it.

The Sun's surface is constantly moving and changing.

Climate change from outside Earth is known as external climate forcing. Internal climate forcing, on the other hand, takes place within the climate system itself.

An asteroid like this can hit Earth at over 3,200kph.

Objects from space

Very occasionally, huge chunks of rock from space, called asteroids, strike Earth. The force of a large impact can send enough dust up into the atmosphere to block or scatter the Sun's rays.

As fewer rays reach the planet, temperatures drop and rapid climate change begins. Some experts think an asteroid impact like the one shown here caused the dinosaurs to die out 65 million years ago.

INTERNET LINKS

For a link to the following website, go to **www.usborne-quicklinks.com** and enter the keywords "science weather".

Websites 1 and 2 See how past events may have affected the world's climate.

205-140 mya
Jurassic period
Early: hot, wet, steamy climates worldwide.
Late: climate changes as Pangaea split continues.

140-65 mya
Cretaceous period
Early: cool climates.
Late: world becomes warmer than today.

65-1.8 mya
Tertiary period
Early Tertiary warm, but by 2.5 mya glaciers cover most of land from Tropics to poles.

1.8 mya - present day
Quaternary period
Climate switches between ice ages and warmer intervals called interglacials.

21st century
Global temperatures have risen by 0.5°C in 100 years. Many scientists predict rapid and widespread climate change. Even so, Earth is cooler today than throughout most of its history.

Gradual climate change

Some kinds of climate change take place very slowly, over millions of years. These changes are too gradual for us to see in action, but we know they have happened in the past.

Astronomical theory

Changes in Earth's path around the Sun, called its orbit, and the way it tilts towards the Sun play an important role in long-term climate change. This is known as the astronomical theory of climate change or the Milankovitch theory.

Changing orbit

The shape of Earth's orbit, called its eccentricity, varies over time. It changes from almost circular to an oval or elliptical shape and back again every 100,000 years.

The nearer Earth orbits to the Sun, the hotter it becomes. The more time it spends far from the Sun, the more it cools down.

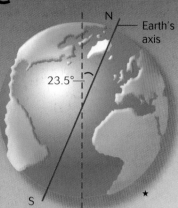

N — Earth's axis

23.5°

S ★

Earth's climate is affected by the way the planet is tilted. Its angle is currently about 23.5°.

Tilting and wobbling

Earth isn't completely upright as it orbits the Sun: its axis, an imaginary line down through its middle, tilts to one side. You may have noticed globes are made like this, too. This tilt is what causes the seasons. Earth's tilt changes between about 22° and 25° every 41,000 years, as it "wobbles" on its axis.

The tilt and wobble affect the amount of difference between the seasons. Combined with variations in Earth's orbit, this heats and cools Earth's climate as a whole. Ice ages begin when summers become cooler and ice builds up throughout the year.

Right now, Earth's orbit is almost circular.

When the orbit becomes oval, Earth travels further from the Sun and starts to cool down.

INTERNET LINKS

For links to the following websites, go to **www.usborne-quicklinks.com** and enter the keywords "science weather".

Website 1 Animations that show how Earth's orbit contributes to climate change.

Website 2 Online explorations of plate tectonics, the Earth's anatomy and the Ice Ages.

Inside the Earth

Earth's core is thought to be made of churning, molten iron, with a solid inner part. The swirling motion of the iron creates both the magnetic poles and a natural magnetic field around the planet. This deflects deadly rays from the Sun.

Every 250,000 years or so the magnetic field weakens to almost nothing, then rebuilds with the poles reversed. The weakened field allows more radiation to heat the upper atmosphere. Anything that has an effect on the atmosphere affects the whole climate system.

Earth's magnetic field (shown in blue) creates a shield (shown in purple) against harmful rays.

Earth

Magnetic field

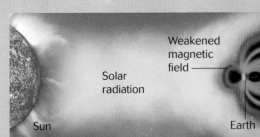

Weakened magnetic field

Solar radiation

Sun

Earth

As more radiation reaches Earth, climate changes. The upper atmosphere heats up, which affects the other layers too.

Moving plates

Climate change over very long timescales is also caused by shifting plates – the huge slabs of rock which make up Earth's outer layer. Plates move very slowly over time, carrying the continents with them.

In the past, many places had a different climate because they were not in the same position on Earth as they are today. For instance, Africa was further south, and so was much cooler.

Pangaea

200 million years ago, scientists believe there was one huge continent called Pangaea.

Pangaea slowly broke up and several smaller continents began to form.

Gradually, the continents drifted to their present-day positions.

Cooling Earth

Since there are many things that can cause Earth's climate to heat up, it needs ways of keeping cool. The land, air and oceans all play a part in this.

Plates and mountains

Earth's outer layer is split into huge slabs called plates, which move around very slowly. Wherever plates collide, mountain ranges build up. As these grow, lots of rock is exposed to the air. Scientists think that this process has helped cause global cooling over many millions of years, by removing carbon dioxide from the air.

Carbon dioxide dissolved in rainwater very slowly breaks down rocks. This creates chemicals called bicarbonates, which contain carbon.

Bicarbonates are washed down to the sea by rivers, where they sink to the ocean bed with their carbon still locked inside.

Cooling down with aerosols

You may know that aerosols come in cans, but they exist in the natural world too, as tiny solid or liquid particles in the air. Aerosols are formed by gases mixing in the atmosphere, or by dust, sand, dirt or pollen rising from the ground. They cool the Earth by reducing the amount of sunlight that reaches it.

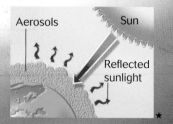

Aerosol particles help to form clouds, which reflect sunlight back into space. This makes the atmosphere cooler.

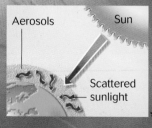

The tiny particles in aerosols can also scatter sunlight in the atmosphere, so less of it reaches Earth's surface.

Where do aerosols come from?

Natural aerosols come from many sources, including forest fires and dust from rocks, soil and volcanic eruptions. Volcanoes give out aerosols called sulphate particles, and a gas called sulphur dioxide.

Sulphur dioxide mixes with water vapour in the air to produce sulphuric acid aerosols. These disappear from the lower atmosphere after a few days, but they can remain in the stratosphere for up to two years.

Today, natural aerosols in the atmosphere combat the effects of global warming, but only a little. In the past, there was enough volcanic activity for them to make a big difference.

Volcanoes give out tiny particles and gases, which gather in the atmosphere and cause the climate to cool temporarily.

Dust storms

Dust from volcanoes and dry land affects how air flows and how heat is moved around. Dust storms are sheets or columns of dust, which whirl up into the atmosphere over deserts. They can block out sunlight and lower land surface temperatures.

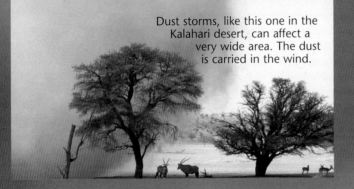

Dust storms, like this one in the Kalahari desert, can affect a very wide area. The dust is carried in the wind.

INTERNET LINKS

For links to the following websites, go to **www.usborne-quicklinks.com** and enter the keywords "science weather".

Website 1 Read about the uncertainties surrounding the effects of aerosols, with helpful diagrams.

Website 2 See how volcanoes are linked to climate change.

Carbon dioxide and oceans

Carbon dioxide dissolves in water, so the oceans play an important part in removing it from the atmosphere. In fact, they may remove up to 85% of the carbon dioxide that we're releasing into the air, and so help to reduce global warming. It's likely that the oceans have had this cooling effect on Earth in the past, too.

Carbon dioxide

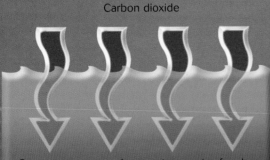

Ocean water can soak up vast amounts of carbon dioxide. It does this best at low temperatures.

Ocean gateways

Ocean currents flow between the continents in natural channels called ocean gateways. As continents drift apart over time, the positions of these gateways change, and the currents flow in different ways.

Antarctica's icy climate is thought to have been caused by the continents drifting apart. Scientists think that around 50 million years ago, a new gateway opened up between South America and Antarctica.

275 million years ago, Antarctica was joined to Australia, Africa, India and South America.

Over millions of years, they split apart. All but Antarctica drifted to warmer areas.

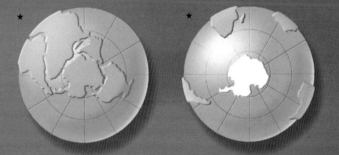

After the gateway had opened, a circumpolar current formed, flowing all around the southern oceans. Antarctica became cut off from the flow of warm water from the equator, and its temperature started to drop. Eventually, it became the icy land it is today.

Antarctic circumpolar and equatorial currents

50 million years ago, the circumpolar current allowed temperature differences between the equator and the South Pole to increase.

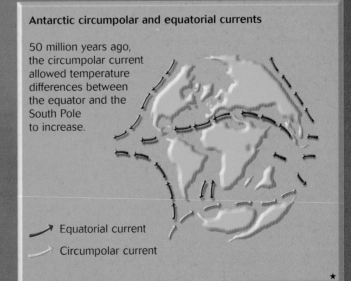

↗ Equatorial current

〰 Circumpolar current

Icy Earth

For most of its long history, Earth has had no permanent ice, but sometimes it has lots. At the moment, not much of Earth is covered in ice. When temperatures drop and permanent ice spreads to more places, it's called an ice age.

Mammoths like this lived during the last ice age. Their shaggy fur protected them from the cold.

Growing glaciers

During an ice age, huge masses of ice called glaciers cover a lot of the land. Glaciers are always growing and shrinking. An ice age is thought to occur when the glaciers are at their largest.

Between each ice age, there is a shorter, warmer period called an interglacial. We have been in an interglacial for about the last 10,000 years. An ice age followed by an interglacial usually lasts about 100,000 years.

What's in a name?

Most major ice ages happened during three very long periods in the last 800 million years. Confusingly, you may see these three periods referred to as "ice ages" too. Also, the last ice age, which peaked 18,000 years ago, is often called "The Ice Age", as if it were the only one.

See for yourself

Ice melts most quickly at its edges, where it is thinner. That's why glaciers start to melt at their lower ends. Next time you eat a bowl of ice cream, look at how it melts. It will start to turn soft at the edges, even if it's still firm in the middle.

Alpine glaciers

Alpine glaciers form in high places and flow slowly downhill, like a solid river. They cut through the landscape, carving out U-shaped valleys as they sweep rocks and soil along with them.

1. At the glacier's top, called the accumulation zone, snow collects and becomes packed down into ice.

2. The ice becomes so heavy that it starts to move slowly down the mountain.

3. As the ice gradually flows downhill, it becomes warmer.

4. At the glacier's lower end, the ice melts.

5. The pure, icy cold meltwater flows into streams and rivers.

Alpine glaciers such as the one below began to form thousands of years ago. They used to cover much more of Earth than they do today.

Continental glaciers

Continental glaciers, or ice sheets, are made of snow which falls in the middle of a land mass and gets squashed into dense ice. When it becomes heavy enough, the ice sheet starts to spread slowly outwards, until it reaches the coast. There, it forms an ice shelf, a thick, frozen ledge which floats on the sea.

Earth currently has two ice sheets – one covers Antarctica, the other covers Greenland. The bigger these ice sheets are, the more sunlight they reflect away from Earth. When they shrink and reflect less, Earth warms up.

Antarctica's ice sheet covers over 13.5 million km².

Greenland's ice sheet is roughly 1/7 the size of Antarctica.

This is the front end of the glacier, called the snout. It is surrounded by meltwater.

Melting and breaking

A glacier starts to shrink if the amount of ice that melts from it is more than the snow that falls on it. Satellite data shows that in the last 30 years, snow cover in the northern hemisphere has decreased by about a tenth, and glaciers are shrinking. This is probably because of global warming.

The speed at which ice shelves break up to form huge, floating chunks called icebergs may also be a sign of climate change.

As Earth becomes warmer, more icebergs, like this one in Antarctica, start to break off from ice shelves.

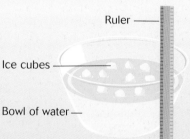

Looking at an ice shelf

Climatologists use satellite images to look at ice shelves. The images below show Antarctica's Larsen B shelf, which broke up unusually quickly in early 2002. The area within the red outline is made up of broken ice spreading into the sea.

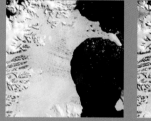

Larsen B ice shelf in January 2002.

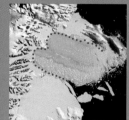

By March 2002, it had partly broken up.

See for yourself

Many people think that melting sea ice – frozen seawater – will raise sea levels. But this is a myth: only melting land ice, from glaciers, can do this. Here's a way to show why.

1. Half-fill a bowl with cold water. Add about ten ice cubes, and measure the water level with a ruler.

Ruler

Ice cubes

Bowl of water

★

2. Wait for the ice cubes to melt, but don't heat the water.

3. Once the ice has melted, measure the water level again. It won't be any higher, because the water took up more space as ice than as a liquid.

Earth's icy past

The world as we know it was very different during the last ice age, which was at its peak around 18,000 years ago. At that time, so much water was frozen into ice sheets on land that the sea level fell by about 90m. Much more dry land was exposed: North America was joined to Asia, for instance. This had an effect on global climate, too.

Ice cover during the last ice age

Key

Areas covered in ice

Frozen lands where animals could live

Dry land and desert

Back then, over 30% of Earth was covered in ice sheets. Today, ice covers only around 10% of it.

During the last ice age, there was less watery weather, because clouds form more slowly at low temperatures. This led to the creation of more grasslands, and eventually deserts, in the areas which were not covered by ice.

The Little Ice Age

Between around 1450 and 1850, the climate suddenly became cooler. It became known as the Little Ice Age. At this time, glaciers spread, sea ice grew, and most areas were at least one degree colder than today. No one knows what caused this climate change.

During the Little Ice Age, the River Thames in London regularly froze over. Markets like this were held on the ice.

A frozen future?

Many experts think that the current interglacial period is coming to an end and the next ice age is overdue. An increase in global warming may delay this, but it probably won't prevent it.

INTERNET LINKS

For links to the following websites, go to **www.usborne-quicklinks.com** and enter the keywords "science weather".

Website 1 See how ice and snow can tell us about climate change.

Website 2 Watch an animation of the retreat of ice sheets from North America.

Website 3 Learn the ancient secrets locked inside Antarctica's ice sheet.

Climate and living things

Animals and plants have adapted to survive in almost every climate on Earth. But for people to live comfortably in extremely hot, cold, dry or wet environments, we have to alter our lifestyles, clothes and buildings to cope with the conditions.

Hot and cold

The most difficult climates for people to live in are extreme heat and extreme cold. Some parts of the world, such as Antarctica, are too cold for people to live in all the time. Many people live in the frozen tundra, though.

Living in the desert

People who live in deserts have to deal with fierce temperatures and an unfriendly landscape. They wear long, light robes and scarves to protect them from sandstorms and the Sun.

This is a camel driver in the deserts of Morocco, North Africa.

People who live in the tundra drink reindeer milk, which this woman is carrying in frozen blocks. The rich milk helps to give them the energy they need to stay warm.

Building design

Buildings can be designed to suit a particular climate. For example, many houses in wet, tropical climates are built on stilts, so they don't become flooded during storms. Some houses in icy climates have steep roofs, so snow slides off them instead of piling up on top.

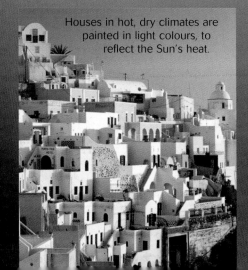

Houses in hot, dry climates are painted in light colours, to reflect the Sun's heat.

Camels can easily deal with extreme desert climates. For example, they shed their fur in the hottest season to keep cool.

Unlike polar animals, a camel's body fat is stored in one place – the hump. If it were spread over their whole body, it would make them too hot.

The human body

Our bodies can adjust to changing temperatures. In hot weather, we sweat. As the moisture evaporates from the skin's surface, it removes heat. Also, blood vessels under the skin widen, taking hot blood nearer the surface. This is what makes us turn red when we feel hot.

In cold weather, our blood vessels narrow, so less heat is lost. We may start to shiver, which warms us up.

Climate and animals

Animals can survive in even more climates than humans, because they have many ways of adapting to extreme temperatures.

Keeping warm

Mammals, such as dogs, can keep warm in cold weather by fluffing out their fur. This traps heat next to their skin. Animals that live in very cold climates, such as polar bears, often have a blanket-like layer of fat under their skin to keep them warm.

Cooling down

Most animals can't sweat, but they have other ways to cool down. Some pant to lose heat, and others lick themselves. The spit acts like sweat, cooling their skin as it evaporates. Some, like the camels above, have specially adapted bodies.

Mammals lose lots of heat through their ears, so, like most Arctic animals, this Arctic fox has small ears.

INTERNET LINKS

For links to the following websites, go to **www.usborne-quicklinks.com** and enter the keywords "science weather".

Website 1 Lots of information, with pictures, about how living things are adapted to climates in different biomes.

Website 2 Explore a desert and survive.

People and their needs

Earth's human population is growing all the time. The more of us there are, the more fuel, food, water and land we need. If we have these things, we don't tend to question where they come from, or what will happen if they run out. In many cases, it's the climate system that suffers from our growing needs.

Population growth

Earth's population is currently around 6.3 billion. Experts predict that it's likely to reach at least 12 billion before it stops growing. All of these people need food and fuel, but it could be difficult to provide them without harming the land and atmosphere.

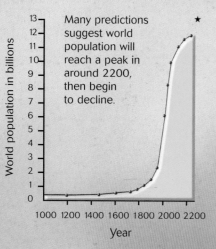

World population since the year 1000

Many predictions suggest world population will reach a peak in around 2200, then begin to decline.

(Graph: Y-axis labelled "World population in billions" from 0 to 13; X-axis labelled "Year" from 1000 to 2200)

Sources of power

Most of the power we need for homes, vehicles and industry comes from burning fossil fuels, such as coal and oil, to generate electricity. This releases the greenhouse gas carbon dioxide, which has been stored in the fuels for millions of years.

Developing world

In wealthy, developed countries, people's comfortable lifestyles require lots of fuel to make energy.

Many developing countries once had very low energy and fuel use. But now, wealthier lifestyles and energy needs are spreading around the world. Using lots of energy can take a terrible toll on the atmosphere.

Using vehicles makes life easier, but it pollutes the atmosphere.

Making space

As populations grow, they take up more space. Farmers also need space for growing crops and grazing animals, both to feed their own people and to sell to other countries.

In some countries, forests are burned down to make room for farmland. The trees release carbon dioxide as they burn, just like fossil fuels.

The way we live generates lots of rubbish. It's taken to vast dumps called landfills, where it rots and gives off methane.

Unnatural gases

Humans are harming the atmosphere with more than just natural gases. CFCs (chlorofluorocarbons) are man-made gases which damage the protective ozone layer in the atmosphere. CFCs were first used as coolants in refrigerators, and in air conditioners and aerosol cans.

A number of international agreements have been put in place to cut down on our use of CFCs. But old used fridges which contain CFCs are still sold to developing countries.

These men in the Madhupur forest, Bangladesh, are burning down trees to make space for farmland.

Facts and figures

Developed countries, such as the UK and USA, emit the most greenhouse gases because they burn more fuel per person than anywhere else. All these countries are trying to reduce their emissions in some way, but it's a slow process.

The table below showsXthe percoJtage of CO_2 eissionsøöer country in 2001. France is separate from the rest of the EU:³ecauseA¥ts emissions aŠ very d fferent.

Country	Average carbon use per person (tonnes)	% of world carbon dioxide emissions
India	0.26	4.2
Brazil	0.52	1.4
China	0.58	11.4
Canada	1.05	2.2
France	1.88	1.7
Russia	1.97	12.6
EU	2.49	14.6
Japan	2.77	5.4
USA	4.44	25.2
Australia	4.73	1.4
Rest of world	0.57	21.6

Source: BP statistical review of world energy, 2001

The Kyoto Protocol

Most governments are worried about rising greenhouse gas levels in the atmosphere, and the effect they may have.

In 1997, representatives from 160 countries met in Kyoto, Japan to create the Kyoto Protocol. This is an agreement to reduce greenhouse gas emissions to 5.2% less than 1990 levels by the year 2012.

Ordinary people as well as government officials shared their feelings in Kyoto.

Kyoto isn't a perfect solution, though. Many environmental campaigners say the actions it suggests are not enough. Developing countries were left out of the agreement, despite their rising carbon dioxide emissions. Some major countries pulled out because they thought it would cost too much to lower their emissions.

INTERNET LINKS

For links to the following websites, go to **www.usborne-quicklinks.com** and enter the keywords "science weather".

Website 1 A good look at the world's main polluters and what they are doing.

Website 2 Watch the world's population grow on a population clock.

Website 3 Find out more about the Kyoto Protocol, with online activities.

What we can do

There are still lots of uncertainties and arguments about the role we play in causing climate change, but we can still do something about it. Governments, businesses and individuals can all make a difference to Earth's future.

★ This symbol means "recycled". Recycling uses up fewer natural resources, so the land and atmosphere are affected less.

Alternative energy

Alternative energy sources, such as wind, waves and solar power from the Sun, are cleaner than fossil fuels because they don't emit greenhouse gases. They are also called renewable energy sources, because they will never be used up. If governments invest in alternative energy, fossil fuel emissions will decrease. Below, you can see the main types of alternative energy.

The Sun's energy is turned into power by these solar panels. But not many places have enough sunlight throughout the year to meet all their energy needs in this way.

Natural heat from beneath Earth's surface is the source of geothermal energy. Only a few countries, such as Iceland, have access to heat near the surface which they can use.

In hydroelectric power stations, flowing water from a dam turns huge, rotating turbines in order to generate power.

Nuclear power is generated by breaking apart radioactive substances, such as uranium. Many people think it's dangerous.

Big solutions...

At the moment, only about 4.5% of the world's power comes from renewable sources. Many countries are still very dependent on fossil fuel, so changing over takes time, money and effort. But world leaders are trying to do it, because they know fossil fuels will eventually run out.

...and little solutions

It's not just up to governments to make a difference. Most people would like to help prevent global warming, but some think their small actions won't be enough. Many are worried that it will be too difficult or expensive for them.

In fact, there are lots of ways to help that barely affect your lifestyle. A good way to start is by using less energy.

Wind and wave power are widely used around coastlines. They use the motion of the sea or the air to move huge turbines, powering a generator.

Saving energy

Whenever we use electricity, power plants work to create the energy we need, and they add more greenhouse gases to the air. There are lots of things we can do to reduce the energy we use. These include:

• switching off lights when leaving a room

Energy-efficient lightbulbs like this one last much longer than ordinary bulbs but use up to 80% less energy.

• turning off a TV, stereo, video or DVD player completely when you've finished using it instead of leaving it in "standby" mode

• turning down your central heating thermostat by as little as one degree

INTERNET LINKS

For links to the following websites, go to **www.usborne-quicklinks.com** and enter the keywords "science weather".

Website 1 Take a trip to Recycle City, where you can learn how to help Earth.

Website 2 Lots of renewable energy information, with a game too.

Cleaning up the roads

All over the world, cars and other vehicles are using huge amounts of fuel. Most cars pollute the air and release a lot of carbon dioxide, but people won't give up driving because cars are so useful.

Instead, we need to cut their emissions. Many car manufacturers are developing hydrogen-powered cars to create a cleaner future. Instead of petrol, these use fuel cells, which turn hydrogen gas and air into electricity.

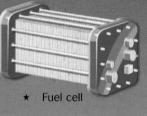

Inside a fuel cell, hydrogen and oxygen combine to generate power. The only waste product is warm water.

★ Fuel cell

This is the Toyota FCHV-4, a car powered by hydrogen fuel cells. With enough investment, cars like this will be widely used in the future.

Recycle and re-use

Recycling or re-using old items helps to reduce landfills, which are harmful sources of methane. Making recycled products often uses less energy than making new ones, because you need fewer raw materials.

Glass, plastic, drinks cans, paper and even mobile phones can all be recycled. Grocery bags and clothes can be re-used, and food waste can be turned into compost.

See for yourself

Many electrical appliances have symbols or letters which tell you how efficiently they use energy. Try checking the appliances in your home to see if they have a rating – an "A" is the most efficient. At school, you can check office equipment, such as computers and photocopiers.

Look for an Energy star© on a computer or a monitor.

The future

If we don't stop adding to the greenhouse effect, Earth will continue to heat up. Even a small rise in its average temperature makes a huge difference. We're already hearing reports of extreme weather, shifting seasons and melting land ice, which seem to be caused by global warming. Many climate experts warn that things will become much worse in the future.

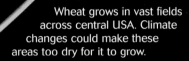

Wheat grows in vast fields across central USA. Climate changes could make these areas too dry for it to grow.

Making predictions

Most predictions about future climates are based on computer-generated climate models called general circulation models, or GCMs. These look at how the atmosphere and oceans work together. Changes to these parts of the climate system have the greatest effect on the world as a whole.

How accurate?

Climate models are like a foggy crystal ball – they give us a glimpse of the future, but it's up to us to interpret what they show. Even the most sophisticated computers can't produce totally accurate models, but many agree that certain things, such as floods, are very likely to happen.

Severe flooding is already causing problems in many countries. Most climate models predict even more in the future.

Wetter and drier

While some places are being drenched by heavy rain and floods, areas that are already semi-dry will suffer drought as the ground bakes in the heat. In extreme cases, the land will turn to desert.

Countries with temperate climates are likely to suffer both floods and drought at different times of year.

Drought is causing this watering hole in New Mexico to evaporate away, and animals and plants are dying as a result.

More hurricanes

Hurricanes form over oceans that are warmer than 26°C. If more parts of the ocean reach this temperature, hurricanes may cause chaos in places where they have rarely been seen before. Some experts also predict that hurricanes in the next century may be up to 20% stronger than today.

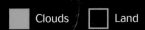
Clouds ☐ Land

This radar image shows a hurricane approaching Florida. Hurricanes are already common in this area but, in future, even more places will come under threat.

Rising sea levels

Sea levels are thought to have risen between 10 and 25cm in the last 100 years. If global warming continues at the current rate, they will rise even further. As the sea level rises around islands and coastlines, the waves wear away the land. Some places could eventually disappear beneath the sea.

North America's threatened coastline ★

The heavily populated areas marked in purple are all threatened by rising sea levels.

The future of El Niño

The four strongest El Niños (see page 35) in the last hundred years have all occurred since the 1980s. If this trend continues, the effects could be devastating. Floods will increase in the Americas, and parts of India and Australia will be paralyzed by drought. Fortunately, computer models and records of past El Niños are making future events easier to predict and prepare for.

INTERNET LINKS

For links to the following websites, go to **www.usborne-quicklinks.com** and enter the keywords "science weather".

Website 1 Take an interactive look at extreme weather and climate change around the world.

Website 2 Problems of predicting future climate change and how global warming may affect people and the environment.

Many types of drugs are made using chemicals extracted from plants. Without Earth's variety of living things, some diseases would be impossible to treat.

Changing climate zones

As temperature and rainfall patterns shift, conditions in climate zones will change too. Living things that are perfectly suited to the climate in which they live may die out if they can't adapt rapidly enough to the new conditions. Loss of species will greatly reduce biodiversity – Earth's amazing variety of life.

Nature's secrets

Loss of biodiversity is bad for all living things, including people. For example, plants from all over the world are used to treat serious illnesses. There could still be hundreds of natural remedies that we don't yet know about.

If the variety of living things decreases, some of these treasures of nature could be lost for ever.

Food chains

All animals eat or are eaten by other living things. This is called a food chain. If one thing in the food chain is harmed, the other parts of it suffer too.

An Antarctic food chain

Tiny sea creatures, called krill

Penguins eat krill.

Seals eat penguins.

Healthy corals are colourful but the pale parts of these corals show that they are dying. Corals eat tiny creatures called algae, but warmer seas are killing the algae.

INTERNET LINKS

For links to the following websites, go to **www.usborne-quicklinks.com** and enter the keywords "science weather".

Website 1 Explore the consequences of global warming and what you can do.

Website 2 A clickable map showing where climate change is hitting hardest.

Global warming is reducing the numbers of krill. So, more penguins starve and die. Fewer penguins means less food for seals.

Plant problems

Plants take in carbon dioxide (CO_2) when they make food for themselves. Raised carbon dioxide levels and higher temperatures cause more plants to grow, but these conditions can be harmful too.

More carbon dioxide and warmer weather makes plants thrive. But carbon dioxide also adds to global warming and this is bad for plants.

★ — CO_2

If soil dries out in the heat, plants cannot take in enough water, so they die. Their stored carbon is released, adding to global warming even further.

★ — CO_2

Carbon dioxide from the ground

Rising global temperatures may cause a massive release of carbon dioxide from soil in the frozen tundra. This happens as permafrost melts, and peat – partially decomposed animal and plant remains – is broken down. The carbon dioxide this releases will also add to global warming, causing even more permafrost to melt.

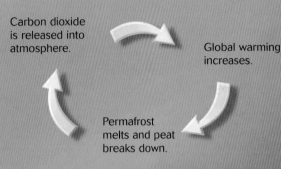

Carbon dioxide is released into atmosphere.

Global warming increases.

Permafrost melts and peat breaks down.

Health hazards

Some effects of global warming, such as floods, threaten food supplies and clean water. Global warming has a direct effect on people's health too. For instance, more people will die from heat stroke during heat waves.

Also, more areas will be warm enough for germs and harmful insects to thrive. This will lead to more widespread outbreaks of insect-carried diseases.

See for yourself

Predictions of future climate change are based on the information available to us, but this changes all the time. You can keep up with current climate news in newspapers, on TV, or on the Internet.

This mosquito is feeding on blood. Warm conditions are ideal for these bloodsuckers, but bad news for people who catch malaria from their bite.

Disagreement

Not everyone agrees that climate change is a threat to our future –
in fact, some experts think we've got it all completely wrong.
A person who disagrees with a popular view is called a sceptic.
Many sceptics are worried that global warming is being treated
as a much bigger problem than it might turn out to be.

Lots of uncertainties

The climate system is very complex, so
predicting climate change is like rolling
dozens of dice: there are so many
possible results, you can't be sure what
you'll get. Feedback systems cause the
most doubt because their effects can
vary so much. Here's an example:

In this negative feedback system, global warming
affects sea level and temperature, which in turn
have an effect on global warming.

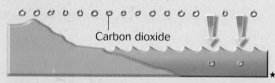
Carbon dioxide

Oceans and tiny creatures which live in them absorb
carbon dioxide (CO_2) from the atmosphere.

If global warming causes sea levels to rise, there
will be more water to absorb CO_2.

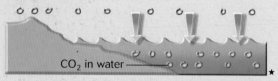
CO_2 in water

The more CO_2 is removed from the atmosphere, the
more it will help reduce further global warming.

This seems straightforward, but it's
not. Global warming also raises ocean
temperatures, making them worse at
absorbing CO_2. Also, fuel and forest
burning could add enough extra CO_2 to
the air to partly cancel out the oceans'
effect. Many parts of the climate system
are as unpredictable as this, so no one
can be sure how it will change.

Varying timescales

Most experts agree that you
must study weather patterns
for many years to get an
idea of a "normal" climate.
Sceptics say that, in
many cases, what
seems like climate change
might actually be normal
conditions, if you look at them
on a much longer timescale.

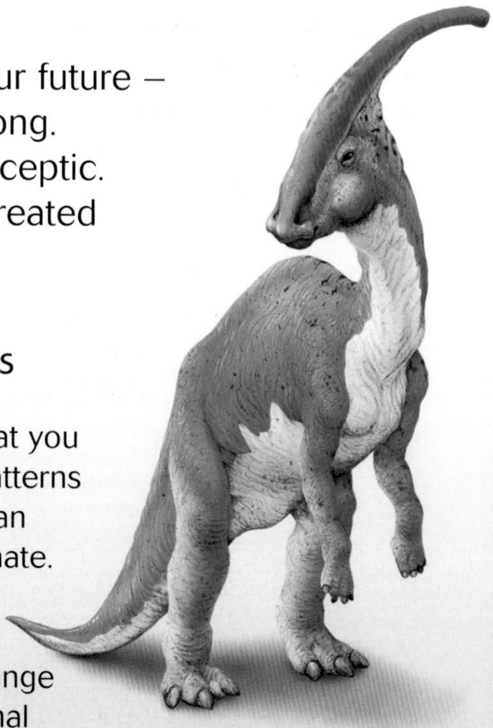

Earth's overall temperature
is high compared to a
hundred years ago. But
compared to the time of
the dinosaurs, it's quite low.

Hot or not?

Global temperature readings are normally taken at
ground or surface level, but features of the landscape
can affect them. For instance, cocoons of heat called
heat islands, which develop around cities, can make
temperatures seem higher than they should.

Example of a heat island

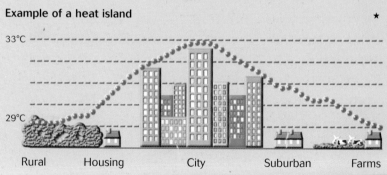

| Rural | Housing | City | Suburban | Farms |

If a surface temperature reading were taken near the city, it would give the
impression that this whole area has a warmer climate than it really does.

According to some sceptics, tree rings and ice
core samples, which aren't affected by small-scale
heat pockets like this, show that Earth's average
temperature has actually barely risen since the 1940s.

Mistaken models

Climate models predict that if more greenhouse gases are heating Earth, this heating should carry on throughout the troposphere as well. Weather balloons and satellite readings show that this isn't happening, so sceptics argue that the models could be wrong about other things too.

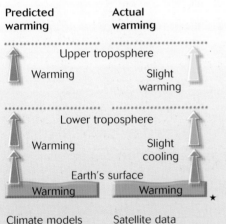

Predicted warming	Actual warming
Upper troposphere	
Warming	Slight warming
Lower troposphere	
Warming	Slight cooling
Earth's surface	
Warming	Warming ★

Climate models predict that raised temperatures at Earth's surface should heat the lower troposphere.

Satellite data reveals the lower troposphere isn't warming. The climate models seem to be wrong.

Out of our hands

Climate is influenced by many things which we can't control, such as changes in energy from the Sun. Some sceptics believe that even if dramatic climate change is taking place, it could have entirely natural causes and so we can't do anything about it.

Many climate models show that climate change will make hurricanes fiercer in the future. Some sceptics say there will be fewer, milder hurricanes. Only time will tell who's correct.

Sceptics in industry

Sometimes, sceptics have a particular reason to challenge popular climate change theories. For instance, they may work for the fossil fuel industry – one of the biggest businesses in the world.

Some parts of the fossil fuel industry want governments to keep investing in fossil fuels, because it would lose money if they stopped. So, it may pay some scientists to show that burning more fossil fuels won't have an effect on future global warming, even though most experts say it will.

Right now, the developed world relies on fossil fuels such as oil, coal and gas for over 80% of its power.

So who's right?

At the moment, most experts agree that global warming is happening, and it's largely due to humans. As research continues and climate models become more sophisticated, we might see more evidence that this isn't the case.

But even if the sceptics are right and we've done no damage so far, we should still do everything we can to avoid causing any in the future.

INTERNET LINKS

For links to the following websites, go to www.usborne-quicklinks.com and enter the keywords "science weather".

Website 1 Explore an online exhibition about climate change then play a game.

Website 2 A detailed and illustrated overview all about global warming.

Climate data

Weather extremes

Highest recorded temperatures

Continent	Temperature (°C)	Location
Africa	58	El Azizia, Libya
America	57	Death Valley, California
Asia	54	Tirat Tsvi, Israel
Australia	53	Cloncurry, Queensland
Europe	50	Seville, Spain
Antarctica	15	Vanda station, Scott Coast

Lowest recorded temperatures

Continent	Temperature (°C)	Location
Antarctica	−89	Vostok
Asia	−68	Oimekon, Siberia
America	−63	Snag, Yukon
Europe	−55	Ust'Shchugor, Russia
Africa	−24	Ilfrane, Morocco
Australia	−22	Charlotte Pass, NSW

Highest average annual rainfalls

Continent	Rainfall (mm)	Area
S. America	11,770	Tututendo, Colombia
Oceania	11,684	Mt. Wai-ale-ale, Hawaii
Asia	11,430	Cherrapunji, India
Africa	10,277	Debundseha, Cameroon
N. America	6,665	Henderson Lake, British Colombia
Europe	4,648	Grkvice, former Yugoslavia
Australia	4,496	Tully, Queensland

Lowest annual recorded rainfalls

Continent	Rainfall (mm)	Location
S. America	0.8	Arica, Chile
Africa	2.5	Wadi Halfa, Sudan
N. America	30.5	Batagues, Mexico
Asia	45.7	Aden, South Yemen
Australia	119.3	Troudaninna
Europe	162.5	Astrakhan, Russia
Oceania	226.0	Puako, Hawaii

Greatest observed rainfalls

Time	Rainfall (mm)	Place
1 minute	31	Unionville, USA
12 minutes	198	Plumb Point, Jamaica
12 hours	1,340	Belouve, Réunion (near Madagascar)
24 hours	1,869	Cilaos, Réunion
1 month	9,299	Cherrapunji, India
1 year	26,459	Cherrapunji, India

Weather records

Greatest snowfall: Tamarac, California, USA, depth of 11.5m, March 1911

Heaviest hailstones: Gopalganj, Bangladesh, up to 1.02kg, April 14 1986

Highest air pressure: Agata, Siberia, 1,083.8 millibars, December 31 1968

Lowest air pressure: West of Guam, Pacific Ocean, 870 millibars, December 10 1979

Strongest wind: Mount Washington, New Hampshire, USA, 372kph, April 12 1934

Synoptic chart symbols

Weather forecasts are compiled by meteorologists. They gather data from sources such as satellites and radars, and plot this information on synoptic charts, like the one shown below. On this type of chart, details of weather conditions are marked in the exact place where the recording was made.

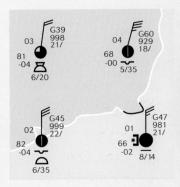

The numbers on each synoptic chart symbol show temperature, visibility, dew point, cloud height and cover, air pressure, and speed of the strongest gust of wind. You can find out more about these numbers on page 58.

Example of a synoptic chart

Present weather symbols

These describe weather conditions such as rain and snow, which may change frequently.

- = Mist
- ≡ Fog
- ⁹ Drizzle
- • Rain
- ⦙ Continuous moderate rain
- ⦙ Continuous heavy rain
- ✳ Snow

- △ Hail
- ⌀ Thunder-storm
- ▽ Shower
- ✳▽ Snow shower
-] Observation in previous hour

Cloud type symbols

You can find out more about these cloud types on page 39.

- ⌒ Cirrus (Ci)
- ⸾ Cirrocumulus (Cc)
- ⸾⸜ Cirrostratus (Cs)
- ∠ Altostratus (As)
- ⌣ Altocumulus (Ac)
- ⧄ Nimbostratus (Ns)
- — Stratus (S)
- ⌒ Stratocumulus (Sc)
- ⌓ Cumulus (Cu)
- ⌂ Cumulonimbus (Cb)

Cloud cover symbols

These show how many oktas (eighths) of the sky are covered by cloud.

- ○ Clear sky
- ◔ 1 okta
- ◑ 2 oktas
- ◑ 3 oktas
- ◕ 4 oktas
- ◕ 5 oktas
- ◕ 6 oktas
- ◕ 7 oktas
- ● 8 oktas
- ⊖ No record
- ⊗ Sky obscured (e.g. by fog)

Wind speed symbols

Wind speed is measured in knots. One knot is equal to 1.85kph.

- ⊙ 0
- —○ 1-2
- ⌐○ 3-7
- ⌐○ 8-12
- ⌐○ 13-17
- ⌐○ 18-22
- ⊫○ 23-27
- ⊫○ 28-32
- ⊫○ 33-37
- ⊫○ 38-42
- ⊫○ 43-47
- ▲○ 48-52

The Beaufort scale

The Beaufort scale was developed in 1805 by Sir Francis Beaufort to estimate wind speed at sea. It was later extended to include exact wind speeds and adapted for use on land. Although rarely used in meteorology, it's still a popular way of measuring wind speed without instruments.

Beaufort number	Kilometres per hour	Wind description
0	Less than 1	Calm
1	1-5	Light air movement
2	6-11	Slight breeze
3	12-19	Gentle breeze
4	20-28	Moderate breeze
5	29-38	Fresh breeze
6	39-49	Strong breeze
7	50-61	Moderate gale (or high wind)
8	62-74	Gale
9	75-88	Strong gale
10	89-102	Storm
11*	103-117	Violent storm
12*	118 or more	Hurricane

* Storms this strong usually only happen at sea.

Lightning types

Name	Description
Ball	Free-floating, bright ball
Forked	Many branches
Pearl necklace	Points of extra brightness
Ribbon	Follows a very bent path
Rocket	Travels very slowly
Sheet	Flashes from behind a cloud
Streak	One main stroke with many smaller branches

Did you know?

Human hair can be used to measure humidity. Hair expands in moist air and shrinks as the air dries out.

Lightning can travel at up to 140,000km per second. At this speed, a rocket would reach the moon in 2.5 seconds.

If the Earth were the size of an apple, the whole atmosphere would be no thicker than the apple's skin.

The largest single hailstone ever recorded fell in Coffeyville, Kansas, USA on September 3 1970. It weighed 750g and was 44.5cm across.

The largest snowflake ever measured fell in Montana, USA on 28 January 1887. It was 38cm across and was 20cm thick.

The light given off by an area of the Sun's surface the size of a postage stamp could light all the rooms in 48 average-sized homes.

There are about 16 million thunderstorms a year throughout the world. About 1,800 storms are raging as you read this.

The most extreme difference between record high and low temperatures is in Verkhoyansk, Siberia. Temperatures there range from −76.8°C to 36.7°C.

Weather discoveries

340BC Greek philosopher Aristotle wrote *Meteolorica*, a book summarizing all weather information known at that time.

1st century BC Hero of Alexander (Greece) was probably the first to discover that air has weight.

1593 Galileo Galilei (Italy) invented the first thermometer: a bulb filled with water.

1611 Johann Kepler (Germany) was the first to identify the six-sided shape of snowflakes.

1643 Evangelista Toricelli (Italy) invented the first barometer.

1654 Grand Duke Ferdinand of Tuscany invented the first sealed thermometer.

1718 Gabriel Daniel Fahrenheit (Germany) devised the Fahrenheit scale (°F) for measuring temperature.

1722 Anders Celsius (Sweden) devised the Celsius or centigrade scale (°C) for measuring temperature.

1742 Benjamin Franklin (USA) invented the lightning conductor.

1783 Horace-Bénédict de Saussure (Switzerland) made the first hygrometer.

1802 Luke Howard (Britain) classified the cloud types.

1805 Admiral Sir Francis Beaufort (Britain) devised the Beaufort scale for measuring wind speed at sea.

1843 Lucien Vidie (France) made the first aneroid barometer.

1846 John Robinson (Britain) invented the cup anemometer.

1851 The first published weather maps were sold to the public (in London).

1856 The first national storm warning system was established in France.

c.1887 Clement Wragge (Australia) was the first person to give hurricanes women's names. Today, an alphabetical list of male and female names is used.

1920s Meteorologists in Norway identified fronts and air masses.

1930 Pavel Molchanov (Russia) launched the first radiosonde.

c.1945 John von Neumann (USA) built the first computer to be used for weather forecasting.

Late 1940s Radars were first used in weather forecasting.

1950s Computer models were first used for weather prediction.

1960 The first weather satellite, TIROS 1, was launched by the USA.

Test yourself

1. Climate is defined as:
A. a predictable pattern of weather over a long period of time
B. how hot or cold somewhere is
C. any kind of unusual weather
(Page 8)

2. The hydrosphere is:
A. water which has turned to ice
B. a layer of rainclouds around Earth
C. all the water on and around Earth
(Page 9)

3. Infrared rays are:
A. high-energy rays
B. low-energy rays
C. visible light rays
(Page 12)

4. Sunspots grow and shrink in cycles of roughly:
A. 5 years
B. 11 years
C. 19 years
(Page 15)

5. The highest layer of the atmosphere is the:
A. troposphere
B. mesosphere
C. exosphere
(Page 16)

6. The ozone layer helps shield Earth from:
A. high-energy rays from the Sun
B. low-energy rays from the Sun
C. meteors
(Page 17)

7. The most commonly occurring greenhouse gas is:
A. carbon dioxide
B. nitrous oxide
C. water vapour
(Page 18)

8. Without the greenhouse effect, temperatures on Earth would be:
A. about 63°C lower
B. more or less the same
C. about 33°C lower
(Page 21)

9. Since the 1940s, Arctic sea ice has decreased by about:
A. 80%
B. 40%
C. 10%
(Page 23)

10. Boreal forest is also known as:
A. savanna
B. tundra
C. taiga
(Page 27)

11. The Gulf Stream is:
A. an ocean current in the northern hemisphere
B. a type of powerful wind
C. an Asian river
(Page 34)

12. Low-level clouds which sit in a layer are called:
A. cumulus
B. stratus
C. cirrostratus
(Page 39)

13. A mixture of rain and snow is:
A. drizzle
B. sleet
C. rime
(Page 40)

14. Low-pressure areas are also known as:
A. monsoons
B. anticyclones
C. depressions
(Page 45)

15. An air mass that forms over hot, dry land is described as:
A. tropical maritime
B. tropical continental
C. polar continental
(Page 48)

16. Hurricanes may form when:
A. warm, moist air rises over the ocean
B. warm, dry air rises over land
C. cool, moist air blows over the ocean
(Page 51)

17. Hygrometers measure:
A. temperature
B. air pressure
C. humidity
(Page 55)

18. Scientists who study weather are called:
A. paleontologists
B. cosmologists
C. meteorologists
(Page 58)

19. On a weather map, this symbol stands for:
A. temperature
B. wind speed
C. humidity
(Page 58)

20. Earth is currently tilted at an angle of about:
A. 21.5°
B. 22.5°
C. 23.5°
(Page 66)

21. Erupting volcanoes produce a gas called:
A. sulphur dioxide
B. ozone
C. sulphur hydroxide
(Page 68)

22. Ocean water is best at absorbing carbon dioxide from the air:
A. at low temperatures
B. at high temperatures
C. at night
(Page 69)

23. An interglacial lasts for about:
A. 10,000 years
B. 50,000 years
C. 100,000 years
(Page 70)

24. Huge, floating chunks of ice in the sea are called:
A. glaciers
B. ice shelves
C. icebergs
(Page 72)

25. Earth's population is currently:
A. just over 3 billion
B. just over 6 billion
C. just over 10 billion
(Page 76)

26. Rotting landfills are a source of:
A. methane
B. oxygen
C. ozone
(Page 77)

27. The 1997 international agreement to reduce greenhouse gas emissions was called:
A. the Tokyo Protocol
B. the Berlin Protocol
C. the Kyoto Protocol
(Page 77)

28. Vehicle exhausts give out:
A. carbon dioxide
B. oxygen
C. methane
(Page 79)

29. The incredible variety of life on Earth is called:
A. natural selection
B. biodiversity
C. biomes
(Page 82)

30. Heat islands are cocoons of heat which form:
A. around cities
B. above the Pacific Ocean
C. on mountaintops
(Page 84)

Glossary

Here are explanations of some of the more difficult weather and climate terms you may have found.

A

aerosols Tiny particles in the air which scatter sunlight. They can be natural or artificial.

air mass A mass of air with the same temperature and *humidity* throughout.

air pressure see *atmospheric pressure*.

altitude Height above sea level.

anemometer Device used to measure wind speed.

anticyclones or **highs** Areas of high *atmospheric pressure*.

astronomical theory of climate change See *Milankovitch theory*.

atmosphere 1. The protective layer of air around Earth that enables plants and animals to live. 2. A layer of gases around any planet.

atmospheric pressure or **air pressure** The force of air pressing down on Earth, measured in millibars (mb).

auroras Patterns of light in the sky, seen around the North Pole (Aurora Borealis) and the South Pole (Aurora Australis).

B

barometer Device used to measure *atmospheric pressure*.

biome An area of Earth with a particular combination of climate, landscape, plants and animals.

boreal forest or **taiga** A wide area of coniferous forest across Northern America, Europe and Asia.

C

carbon sink Part of the landscape that absorbs and stores carbon dioxide from the air. Forests and oceans are carbon sinks.

CFCs (chlorofluorocarbons) Man-made chemicals that are thought to damage the ozone layer.

circumpolar current An ocean *current* that flows without interruption around Antarctica.

climate forcing Any human-made or otherwise artificial changing of the *climate system*.

climate model A construction of past or future climate conditions, created by feeding various types of data into a computer.

climate system The complex relationships between all the features that create climates on Earth. These include the Sun's heat, living things, water, air and ice.

cloud cover The amount of sky seen to be covered by cloud in any given place.

cold front A boundary at the head of a cold *air mass*.

condensation The process of water vapour turning back into liquid water as it cools down. See also *evaporation*, which is the opposite effect.

conduction The transfer of heat from one substance to another, or through a substance. It always takes place from warmer to colder areas.

continent Any of the large land masses into which Earth is split.

continental air mass An *air mass* that forms over land.

continental drift The slow movement of the continents, caused by the shifting of Earth's plates.

convection The way heat energy in liquids or gases is transferred. The warmer part rises and the cooler, denser part sinks.

coriolis effect The deflection of winds from their direct paths between areas of high and low pressure, caused by Earth's rotation.

currents Huge bands of air or water which carry heat around Earth.

cyclone See *lows*.

D

depression Another term for an area of low pressure, or cyclone.

developing country A country where, generally, many people rely on farming: advanced technology is limited and wages are low.

dew Water droplets which form at ground level on cool, wet nights, as temperatures decrease and water vapour in the air condenses.

dew point The temperature at which water vapour turns back into a liquid.

doppler radar A type of radar which can detect the direction and speed of approaching wet weather systems.

E

El Niño A regular reversal of *air pressure* in the tropical Pacific. It brings extreme weather.

equator An imaginary line which runs around the middle of the Earth, dividing it into the northern and southern hemispheres.

equatorial climate A climate which is always hot and wet. Rainforests are found in regions with an equatorial climate.

eras The four main divisions of geological time: Precambrian: 4,600-545 million years ago (mya); Paleozoic: 545-250 mya; Mesozoic: 250-65 mya; Cenozoic: 65 mya to the present day.

evaporation The process of liquid water turning into water vapour, as it is warmed. See also *condensation*, the opposite effect.

exosphere The highest layer of the atmosphere, which merges into space at about 10,000km.

F

feedback system A system where the effect of an action adds to (positive feedback) or decreases (negative feedback) the action itself.

flurry A short, light shower of snow.

fossil fuel A fuel such as coal, oil or natural gas, formed from the fossilized remains of living things.

front A boundary between two *air masses* with different temperatures and *humidity* levels. Fronts can be warm, cold or occluded.

G

geostationary satellite A *satellite* which *orbits* at the same speed as Earth spins, so that it remains fixed over the same point.

geothermal energy Heat energy from underground rocks, used to generate electricity by heating water to steam, which turns turbines.

glaciation 1. The process which forms huge, slowly-moving rivers of ice called glaciers. 2. A period of time during which glaciers on Earth are at their greatest.

global energy system The transfer of heat from places where there is a lot of heat to places where there is less.

global warming A rise in average temperatures around the world which scientists believe to be caused by the *greenhouse effect*.

global winds Winds caused by the movement of air between the *equator* and the poles.

greenhouse effect The trapping of heat by carbon dioxide and other greenhouse gases in Earth's *atmosphere*. An enhanced greenhouse effect is created when more greenhouse gases are added.

H

heat island A "pocket" of heat which forms around a city. It's sometimes visible as a heat haze.

highs See *anticyclones*.

hoar frost Spiky ice crystals which form on objects near the ground.

humidity The amount of moisture in the air.

hurricane A fierce storm caused by warm air rising over the ocean.

hydroelectric power A form of electricity generated by water turning turbines, which are connected to generators.

hydrosphere A collective name for all the water on and around Earth.

hygrometer A device used to measure *humidity*.

I

ice ages Periods of time when parts of Earth are permanently icy.

ice sheets Huge plains of permanent ice. Earth has two ice sheets: one covers Greenland, the other covers Antarctica.

infrared radiation (infrared rays) Low-energy radiation given out by the Sun, or anything hot.

instrumental data Information gathered using specially designed instruments, such as *satellites*.

IPCC (Intergovernmental Panel on Climate Change) A group of scientists from around the world who gather and assess climate data, to produce reports on it.

isobar A line on a weather map connecting points of equal *atmospheric pressure*.

ITCZ (Intertropical Convergence Zone) An huge band of low pressure around the middle of the Earth.

J

jet streams Strong winds concentrated within narrow bands of the atmosphere, at 30° and 60° north and south of the equator.

L

La Niña A period of unusual cooling in the eastern tropical Pacific, which often follows *El Niño*.

lapse rate The rate at which temperatures in the *atmosphere* decrease with height.

local climate The climate of a relatively small region, which differs from that of its surroundings.

local winds Small-scale winds which move and change regularly.

lows Moving areas of low pressure, also known as cyclones or depressions.

M

maritime air mass A humid *air mass* that forms over the ocean.

mass extinctions Six points in Earth's ancient past when huge numbers of living things died out in a relatively short time.

meteorology The science of weather, studied by **meteorologists**.

microclimate A climate within a very small area.

mid-latitudes The areas of the globe which lie roughly halfway between the *equator* and the poles, at about 30° north and south.

Milankovitch theory A theory that attempts to explain long-term patterns of climate change. It focuses on Earth's *orbit* and angle in relation to the Sun.

monsoon The name given to changing winds in the Tropics, which cause a fierce rainy season to begin suddenly.

N

nitrous oxide A powerful greenhouse gas, given out by fertilizers and vehicle exhausts.

nuclear energy The energy held in the nucleus of an atom. Nuclear power comes from this.

O

occluded front or **occlusion** A type of *front* created by a warm front and a cold front joining together.

ocean gateways Gaps between land masses which ocean *currents* flow through.

offshore breeze A breeze that blows from the land out to sea.

okta A measurement of cloud cover, with 8 oktas being the highest.

onshore breeze A breeze that blows from the sea to the land.

orbit 1. To circle around an object. 2. The path taken when orbiting.

oxygen A gas taken in by living things: animals breathe it, and plants take it in through their leaves.

ozone layer The protective layer of ozone gas in the stratosphere, which absorbs harmful ultraviolet radiation from the Sun.

P

paleoclimatology The study of ancient climates. A person who does this is a paleoclimatologist.

Pangaea A single giant land mass that began to split up about 225 million years ago, leading to the formation of the modern continents.

periods Subdivisions of ancient *eras*, each lasting many millions of years.

permafrost Permanently frozen ground, found mainly in the *tundra*.

plates or **tectonic plates** Sections of Earth's outer layer, which move about slowly and fit together like the pieces of a jigsaw.

plate tectonics The study of the movement of Earth's tectonic plates.

polar air mass A cold *air mass* that forms in polar areas.

polar or **semi-stationary front** A huge, permanent *front*, made up of a series of *weather fronts*. There is one polar front in each hemisphere.

polar-orbiting satellite A *satellite* which *orbits* Earth from pole to pole.

precipitation Any kind of wet weather, such as rain or snow.

prevailing winds The most frequent direction of winds in any given area.

proxy data Data which is used when direct evidence for something is not available.

R

radar An instrument used for remote weather detection. It does this by sending out radio waves, which are bounced off objects in the *atmosphere*, such as raindrops and ice crystals.

radiosonde Small device carried by a *weather balloon*. It measures *air pressure*, temperature and *humidity* from the *atmosphere*.

rime Grain-like ice which forms when water drops freeze directly onto a surface.

S

satellite Any object *orbiting* a star or planet. This includes the technological devices orbiting Earth that are used to gather data or receive and transmit radio signals.

sceptic Someone who disagrees with a popular or widely-accepted point of view.

sedimentary rock A type of rock formed when mineral-rich material, called sediment, is deposited, buried and squashed into layers.

smog Air which restricts visibility because it contains pollution.

solar panels Smooth, flat panels which gather energy from the Sun.

solar radiation or **solar energy** Radiation from the Sun, made up of high-energy ultraviolet rays, visible light and low-energy infrared rays.

stratosphere The layer of the atmosphere which contains the ozone layer.

sunspot A small, dark, slightly cooler patch of the Sun's surface.

sustainable, or **alternative, energy** A source of power, such as the wind or Sun, which will never run out.

synoptic chart A chart drawn up by meteorologists, containing detailed information about the weather in the form of symbols.

T

taiga See *boreal forest*.

temperate climate A climate with rainfall all year and temperatures that vary with the seasons.

temperature inversion or **inversion** A situation where temperature increases with height, as layers of warm air are trapped beneath cold air.

thermal A rising column of warm air, which forms over heated land.

tornado A spinning column of air that extends from a cumulonimbus cloud down to the ground.

Tropics The area between the Tropic of Cancer (30°N) and the Tropic of Capricorn (30°S).

trade winds Winds which blow in the Tropics, normally from east to west.

tropical cyclone See *hurricane*.

troposphere The lowest layer of the *atmosphere*, containing 80% of its gases and all of its weather.

tundra A cold biome which exists around the Arctic circle, and is mostly made up of frozen land.

turbulence Any irregular or disturbed flow of air.

typhoon See *hurricane*.

U

ultraviolet radiation (ultraviolet rays) High-energy radiation from the Sun.

upwelling The rising of cold water from the deeper regions of the ocean towards the surface.

V

visibility The maximum distance you can see things ahead of you.

visible light rays The part of solar radiation that allows us to see.

volcano An opening in Earth's surface formed by hot, liquid rock called magma bursting out.

W

warm front A boundary at the head of a warm *air mass*.

water cycle The constant exchange of water between the oceans, atmosphere, land and living things.

weather balloon A balloon which carries a *radiosonde*. The way it rises shows wind speed and direction.

weather front A type of *front* that brings short-term patterns of changeable weather.

westerly winds or **westerlies** Winds which blow from the southwest (in the northern hemisphere) or the northwest (southern hemisphere). They exist at around 60° north and south of the equator.

wind chill The effect of a cold wind making temperatures seem lower than they really are.

Index

93